Diane Harris 1997

Toy Shop

1997 Annual

Foreword

If you're a toy collector, chances are you're familiar with *Toy Shop* magazine. But you may not be aware of just how involved the staff is in bringing you the latest and greatest in the world of toys. I'd like to tell you more about us.

The editorial staff, consisting of myself and associate editor Mike Jacquart, are journalists by trade. But toy collecting has become more than just a subject we write about. It has become part of our daily lives in much more intangible ways.

Because we are immersed in the world of toys more than 40 hours a week, sometimes even our time away from work brings us back to toys.

We attend many toy shows and flea markets, both on work time and for enjoyment on weekends. Our intimate knowledge of values and conditions of toys makes our shopping experiences all the more exciting, whether its at a 1,000-booth show or a neighborhood garage sale. Our excitement at finding a great deal is just as real as yours would be.

Our desks and homes are crowded with characters as diverse as Quasimodo, Humpty Dumpty, Malebolgia, Marvin the Martian, Joe Cool, Princess Leia and Mr. Spock.

In its 10 years of publishing, *Toy Shop* has received a glowing reputation both in the toy collecting world and in the general media. We are often called for quotes for newspaper and magazine articles. Recognized as the definitive source of information on collecting toys, *Toy Shop* has received mention in *USA Today*, the *Chicago Tribune*, *Good Housekeeping* and the *Milwaukee Journal-Sentinel* and on The Discovery Channel.

So, throughout the course of a year, our editors remain involved in toy collecting both inside and outside the office. And when we're not working on *Toy Shop*, we're busy on the next edition of the *Toy Shop Annual* or our popular annual price guide, *Toys & Prices*.

A Little Bigger . . . and Better

The book you have in your hands is the fifth edition of the *Toy Shop Annual Directory*. You'll find much information to access year round in your quest for collectible toys. Responding to your requests, we've kept and expanded listings of nationwide dealers and toy clubs. And, of course, we'll continue to include a calendar of 1997 toy shows to help you plan your outings.

To keep you apprised of the growing trends, values and auction happenings, check out our Market Update and Auction Action stories.

Help us celebrate milestones in the toy industry as well as we honor the 100th birthday of Louis Marx and recognize notable anniversaries in the toy world.

Like any yearbook, we'll take a look back at the previous year. What were the hot toys of 1996? What were the flops? And what does that mean for the future of those toys?

Enjoy this edition of the *Toy Shop Annual*, and use it often. We're sure you'll find it an indispensable source and interesting read.

Thanks to all the contributors who took time to provide interviews for this annual, especially the many national toy companies, dealers, collectors and appraisers; associate editor Mike Jacquart; contributing writers Guy-Michael Grande, Bill Underwood, Beth Wheeler and Jim Trautman; Ross Hubbard and Mary Lou Marshall for cover photography and design; and the advertising staff of Toy Shop *magazine, an essential part of publishing this annual.*

Sharon Korbeck

On the Cover

The Olympics and the presidential election made top news in 1996, but collectors were more interested in what new toys were debuting. That's because the newest toys on the store shelves often see immediate interest and increase in value on the secondary market.

Some of the hottest toys of 1996 took their lead from the top movies on the big screen. *The Hunchback of Notre Dame* and *Independence Day* provided plenty of toys, but will people still be collecting Quasimodo and the *ID4* aliens years from now? And what about late 1996 entries like *Mars Attacks* and *Space Jam*?

On our cover are some of our favorites from the past year. Here's an identification guide to the toys pictured on our front and back covers.

From Applause: *The Hunchback of Notre Dame* PVC figure of Quasimodo, latex gargoyles puppets, Esmeralda Keepsake Doll; **Colorforms:** *The Hunchback of Notre Dame* Play Set; **Ertl:** Die-cast replica of Plymouth Prowler (American Muscle series); **James Industries:** Slinky Dog from *Toy Story;* **Kenner:** 12-inch Classic Collection G.I. Joes; **Lewis Galoob Toys:** Jonny Quest figures, Sky Dancers, DragonFlyz; **Mattel:** Star Trek Barbie and Ken, Cabbage Patch OlympicKids, Hot Wheels; **McFarlane Toys:**13-inch Angela figure (Spawn); **Playmates:** *Star Trek* and Flash Gordon figures; **Radio Flyer:** Little Classic wagon; **Revell-Monogram:** *Star Trek Voyager* and Olympic model kits; **Thinkway Toys:** Talking Buzz Lightyear and smaller *Toy Story* figures; **Trendmasters:** *Independence Day* figures, vehicles and sets; **Tyco Toys:** Matchbox cars and action play sets.

* * * * *

Thanks to the following companies for providing toys for the cover: Applause, Colorforms, Ertl, James Industries, Kenner, Lewis Galoob Toys, Mattel Toys, McFarlane Toys, Playmates, Radio Flyer, Revell-Monogram, Thinkway Toys, Trendmasters, Tyco Toys.

Table of Contents

1997 Toy Shop Annual Directory
Editor: Sharon Korbeck
Associate Editor: Mike Jacquart
Contributing Writers: Guy-Michael Grande, Jim Trautman, Bill Underwood, Beth Wheeler
Cover and page design by Mary Lou Marshall

ISBN 0-87341-472-1. Published in the United States of America by Krause Publications, 700 E. State St., Iola, WI 54990-0001, 715-445-2214, publishers of *Toy Shop* magazine.

Advertiser Index

Grading Your Toys

The Facts About Condition

By Sharon Korbeck

Isn't it funny how that toy you bought in Excellent condition just last month suddenly becomes Near Mint when you're ready to sell it? Judging the condition of a toy is basically in the eye of the beholder — and that often depends on which side of the bargaining table the beholder is on.

Let's face it — judging the condition of toys and placing them on a grading scale is far from a scientific determination. It is basically just a word game — one that you must learn to play, however, if you're aiming to become a serious collector.

Determining the value of toys is a subjective game, but a serious one among collectors. What you will pay one day in New York for an item could vary greatly from the going price in New Orleans. Because of this, price guides and grading scales are not absolutes, but parameters within which to operate when buying or selling. Even slight perceived differences in quality and condition can affect the asking price.

Grading scales vary from dealer to dealer, publication to publication, and collector to collector. Because people collect for different reasons, they have different expectations. Some people seek toys in the best possible condition for investment purposes; others may seek the same toy in any condition for nostalgic reasons.

Some dealers can be quite arbitrary about grading scales, and "Good" condition may not mean the same thing to two dealers. Those differences can warp the market. Keeping a consistent grading system can give dealers a good reputation.

While grading is an imperfect science, there are some constants. Packaging is one of those considerations. Most toys originally packaged on cards or in boxes command higher prices in their new, pristine condition even though they may be more beautiful outside the package. A good example of this is model kits. While the boxes may have beautiful artwork, the model itself can be quite stunning when assembled. Put together Godzilla's Go-Kart, however, and you'll send his price toppling.

What about those collectors who actually enjoy the thrill of playing with the toy and still want to maintain a valuable collectible? Some people buy two — one to have in the box and one to display. It's all a matter of personal preference, and whether or not the toy was purchased as a future investment. That advice may hold true for many items, but few people can afford that luxury with more expensive and scarce items.

If you are lucky enough to own a toy in its original packaging, the condition of that packaging alone greatly affects the toy's value. Does the box exhibit shelf wear? Are the edges worn or lightened in color? Does a blister card have a crease or scratches? Is the plastic bubble perfect? Even subtle imperfections are a big determinant in what a hard core collector will pay.

Individual types of toys may have their own grading criteria, but the consistent link is appearance. Does the toy look used? Is it rusty, dented or scratched? Are there parts or pieces missing?

Toy Shop has adopted the following grading descriptions, and remember, no single grading system will apply to all toys. Some descriptions may not apply to certain toys, depending on how they are categorized, when they were produced, how they were originally packaged and how they are collected today.

MIB (Mint in Box): Just like new in the original package. The box may been opened, but inside packages remain unopened. Blister cards should be intact and unopened. New toys in factory-sealed boxes may command higher prices.

MNP (Mint No Package): A toy in Mint condition but not in its original package.

NM (Near Mint): A toy that looks new in overall appearance, but exhibits very minor wear and does not have the original box. An exception would be a toy that comes in kit form. A kit in Near Mint condition would be expected to have the original box, but the box would display some wear.

EX (Excellent): A toy that is complete and has been played with. Signs of minor wear may be evident, but the toy is very clean and well cared for.

GD (Good): A toy that obviously has been played with and shows general wear overall. Paint chipping is readily apparent. In metal toys, some minor rust may be evident. In sets, some minor pieces may be missing.

Take a look at our grading system and compare it with others you may encounter. The basic tenets of condition and appearance apply, although perhaps at different levels.

Decide what type of collector you want to be — the level of money you are willing to spend, the amount of return (if any) you hope to reap from your investments and the amount of time you will devote to your hobby.

Do you hungrily crave a Marx Merrymakers' Band in Mint condition to add to your collection? Or do you just like the appearance of old tin toys, even if they are in less than Excellent condition? Your purpose and mission as a toy collector will figure into how much you are willing to spend.

So, learn the jargon, get a good eye for subtleties, compare grading scales and decide what it's worth to you. You can turn the subjective field of toy collecting into a rewarding objective for yourself.

How Licensing Affects a Toy's Success

By Jim Trautman

How does Michael Jordan's face end up on a Starting Lineup figure? Who has the right to make toys based on Walt Disney characters?

Long before toys hit store shelves, the legal labyrinth of licensing begins.

To use the name or image of a person or character on a toy or related product, the manufacturer must pay a fee to the license holder for that right. Those fees are negotiated based on the licensor, license seekers, competition and potential profitability of the property. A potentially hot property may command more interest from companies hoping to turn profits. That might sound obvious, but knowing what properties will be hot is often a blind gamble.

Take, for example, *Star Wars* — one of the today's hottest licenses. It wasn't always an obvious winner.

When the movie came out in 1977, 20th Century Fox controlled the license to the *Star Wars* property. But there was little indication that the movie would be a blockbuster, so licensing fees were lower than for the later *Star Wars* films.

In fact, Kenner acquired rights to produce *Star Wars* toys, but didn't rush into production. The first action figures, now known as the Early Bird Kit, didn't appear until after Christmas, 1977.

But, of course, the *Star Wars* property has eventually become a goldmine for Kenner.

In another instance, however, Kenner's acquisition of the *Waterworld* license turned out to be less advantageous. The movie was not a success, and neither were the toys.

Securing a license to a property often amounts to bidding wars between companies willing to spend millions of dollars. In return, the companies anticipate that their toy lines will sell more — based on the strength of the license.

For example, Walt Disney properties are among the hottest licenses to acquire. Toys with a tie to a successful Disney film are expected to be sure-fire winners. That was certainly the case with *The Lion King* and *Toy Story*, but films like *The Hunchback of Notre Dame* were more despairing at the box office and more lingering on toy shelves.

Toy companies can't just rely on money, however, to acquire a license. It isn't always a matter of who can pay more. A company's track record, distribution system and reputation are key considerations.

Image is everything. Walt Disney, as well as other licensors, wants the toys bearing their name to project the proper image. Character depictions are reworked, remolded and repainted until the licensor is happy with the likeness.

A HOT LICENSE — Licensing can turn a good collectible into a great collectible. Harley-Davidson licensing makes this Revell-Monogram model kit more desirable to motorcycle fans.

Keeping the relationship open and healthy between the licensor and licensee is vital. Kenner has been producing Starting Lineup hockey figures for several years. According to Ilene Kent of the National Hockey League (NHL), the license was granted because the figures enhance the image of the league and its players.

The Starting Lineup series also illustrates how complicated the licensing maze can be. Even though the NHL licenses the team uniforms and logos, a separate license is required from the National Hockey League Players Association to reproduce the images of the players.

Kenner's reputation for producing quality action figures is long-standing. But it may be difficult — but not impossible — for a newer or relatively unknown company to acquire a major license.

Two recent exceptions, however, were the selection of Canada's Thinkway Toys as major licensee for the *Toy*

Story property and Trendmasters, which produced *Independence Day* toys.

Beware of Unlicensed Imitations

Toy manufacturers know how valuable licensing can be to a product's success. It is so valuable, in fact, that unlicensed imitations of toys cost the industry millions each year.

Some properties — like *The Hunchback of Notre Dame* — fall into the public domain. That means any company can produce a toy based on that character. Many did, as evidenced by the rash of "knockoff" toys Quasimodo and the other characters in the film.

Using Disney's name, movie logo and character likenesses without permission, however, would be a licensing infringement.

Some companies unable (or unwilling) to secure a license may try to capitalize on the potential success of a character by producing an imitation toy. Imitations often have similar packaging, logos and product design as licensed toys, but they cannot legally carry the licensor's name. Although similar, imitations often look just a little different from originals.

Licensing of Warner Bros.' Space Jam movie may prove a big score for Applause and Playmates.

Sometimes a licensed name becomes so popular that a rash of similarly-named and toys hit the market. For example, when Bandai's Mighty Morphin Power Rangers became hot, scores of similar action figures with names like Mighty Rangers or Super Rangers surfaced. Because they were packaged similarly, the toys were designed to compete with the originals. The hope was that buyers would purchase the imitations, which are often less expensive.

But just as toys come and go, licenses do as well. The length of licensing agreements vary. In some instances, a company may own a license for only one or two years; some are longer.

Based on the success of a license, a company may let a license lapse or try to re-negotiate a longer arrangement.

Not all toys, of course, are produced with licensing agreements. But many are, and often it is that license that can make or break a product line.

Jim Trautman is an eclectic collector who frequently writes for Toy Shop *and other publications. He can be reached at R.R. 1, Orton, Ontario L0N 1N0, Canada, 519-855-6077.*

To Infinity and Beyond . . .

Buzz and Barbie top the Hot Toys of 1996

By Sharon Korbeck

The ruler of toy shelves in 1996 wasn't a Flipper or a Hunchback.

Rather, it was Buzz Lightyear, Barbie and *Independence Day* aliens that battled for the dollars and devotion of eager buyers and collectors.

Even though the general toy buying public may be fickle, collectors *know* what they want. They don't wait for trends. They don't buy on whim (usually). And they often turn their prescience into after-market monetary success.

And that's why some of 1996's best-selling toys may come as no surprise to those intimately involved in toy collecting's secondary market.

The stars at the box office in 1996 were good barometers for toys, in most cases. One of the most anticipated and heavily-promoted children's films of the year, Disney's *The Hunchback of Notre Dame*, disappointed both in theaters and in toy stores. Whether it was timing or subject matter, Quasimodo and friends just couldn't live up to the success of Disney's animated film which preceded it, *Toy Story*.

Almost every major toy manufacturer had a license to produce everything from *Hunchback* plush, dolls, games, puzzles and more. But "*Hunchback* didn't do as well as we thought," said Toys R Us spokesperson Mike Bertic.

Months after the movie had closed in theaters, plenty of *Hunchback* toys remained on shelves, and puppets lingered in Burger King stores. The adorable misfit hunchback may not have charmed collectors initially, but he may be the sleeper character collectible of the year — much like Jack Skellington was for Disney's earlier *Nightmare Before Christmas*.

Another Disney property, *Toy Story*, holds an even brighter future for collectors, and it did well for merchandisers in 1996 as well. When the movie came out in late 1995, there were few, if any, toys based on the film's main characters, Woody the cowboy and astronaut Buzz Lightyear.

But in a surprising coup, the Canadian company Thinkway Toys secured the major toy license for *Toy Story* and began producing action figures and more. The toys were a hit, especially Thinkway's large Talking Buzz Lightyear which proclaimed, "To infinity . . . and beyond!"

Initially, *Toy Story* toys were scooped up by collectors, but by the end of 1996, they could still be found.

Interestingly, buyers weren't just interested in toys marketed with the movie's name and logo but related toys as well.

"*Toy Story* pushed interest in other toys," Bertic said. Classic toys that were depicted in the film — like Playskool's Mr. Potato Head, Ohio Art's Etch-a-Sketch and James Industries' Slinky Dog — also sold well because of the film's popularity.

But the film that blew other movies off the screen (and related toys off the shelves) was *Independence Day* (ID4). "*ID4* did very well. [The toys] were different," Bertic said.

ID4 toys also helped put Trendmasters, manufacturers of the line, on the toy collecting map. The alien figures, with stunning detail and action features, seemed to be favorites. But the human figures seemed to sell a little better, according to a Trendmasters' spokesperson.

Two movies set for late 1996 release were expected to spawn

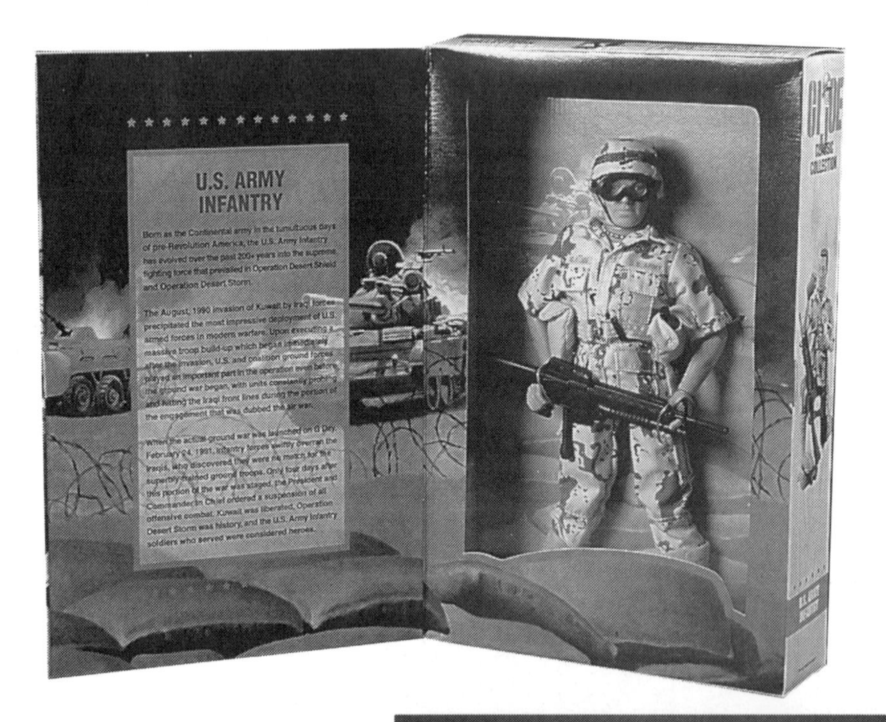

highly-collectible toys.

Warner Brothers' *Space Jam*, starring Michael Jordan and the Looney Tunes, will be complemented by a line of action figures and other toys by major licensee Playmates. A talking Michael Jordan figure is expected to score big with collectors and Jordan fans.

Tim Burton's *Mars Attacks* will be represented on toy shelves by a complete line of figures from Trendmasters.

High-End Barbies Shine

Movie properties may be risky for toy makers, but one licensed property remains good as gold — Barbie.

Mattel's queen of the fashion dolls generally is a big seller for stores, and Bertic said 1996 was no different. "There is a lot more availability [of product] and more collectors," he said. New issues, particularly high-end collector series dolls, sold well.

But some of the regular issue, more generic, Barbie dolls withered on shelves, according to Barbie expert Marcie Melillo.

Melillo, author of *The Ultimate Barbie Doll Book* (Krause Publications, 1996), said, "This is the only time I've seen Barbies marked down and put on clearance in mid-year." That rarely happens, Melillo said.

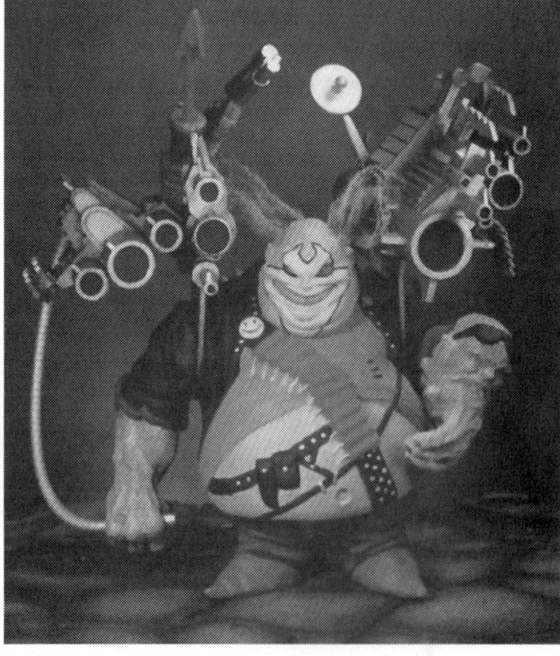

One such doll, she noted, was Flying Hero Barbie, which she called "a complete bust" on shelves. The doll was designed to compete with Bandai's otherworldly Sailor Moon dolls.

Another lower-priced Barbie series that sold well, according to Melillo, was Mattel's American Stories and International series. "They're done well, and they feature inexpensive, attractive costumes," she said.

A set that started heating up early in its release was the attractively-boxed Star Trek Barbie and Ken. The $75 set "should be a smash," according to Melillo. "It's going to cross so many lines," and attract both Barbie and *Star Trek* collectors.

But 1996 was truly the year for high-end collectors' edition and exclusive Barbie dolls to shine. FAO Schwarz's exclusive Statue of Liberty doll, retailing for $75, attracted many buyers with her stunning red-white-and-blue sequinned costume.

Amanda Gronich, spokesperson for FAO Schwarz, said, "This [1996] was our year for collectibles and exclusives. Barbie is bigger than ever. Our Barbie store is growing by leaps and bounds."

One high-end Barbie drawing preliminary mixed reviews was Pink Splendor. The late 1996 release was slated to have a $900 price tag — the highest-priced new Barbie to date.

Some collectors, like Melillo, were predicting a sell-out, but she wasn't sure just who would be buying it.

"Some secondary market dealers are of the mindset that they can pick up any Barbie and resell it for more," Melillo said. That isn't always the case.

The vinyl doll's price tag is in part due to its 24-karat gold thread and Swarovski crystal accessories. Time will tell if collectors will bite.

More Winners of 1996

Other toys that arrived on shelves in 1996 were also big winners. Die-cast vehicles remain hot, fueled by standbys like Mattel's Hot Wheels and Tyco's Matchbox cars.

But according to Gronich, "It was the year of Johnny Lightning." Playing Mantis's die-cast lines have expanded the vintage series. Adults remembering the original 1960s Johnny Lightning cars by Topper Toys may be seeking the new ones as well.

"With more adult buyers," Gronich said, "there's a nostalgia craze." Parents are trying to introduce their children to the toys they played with as kids.

Racing Champions cars, most which feature a NASCAR theme, also were popular, according to Bertic.

In the competitive world of the action figure aisle, dozens of series did battle with McFarlane Toys' Spawn coming out on top. With unique characters and quality production, both kids and collectors can see the value of these pieces.

According to information on Toys R Us's web page, other top selling action figure series of 1996 included Spider-Man, *Star Wars*, *Batman Forever*, and Auto-Morphin Power Rangers.

After many reincarnations, apparently Bandai's Mighty Morphin Power Rangers are still hanging on as favorites on toy shelves. New lines of Zeo Power Rangers were introduced in 1996, creating a whole new series of adventures for the group.

While primarily children are the target buyers for the Power Rangers, adults are actively seeking additions to their *Star Wars* collections. Action figures are especially hot. And both *Star Wars* and puzzle collectors are buying Milton Bradley's 3-D puzzle of the Millennium Falcon.

Figures such as Kenner's Starting Lineups, particularly the basketball's Dream Team players, were also favorites among kids and collectors.

In the board game arena, Monopoly is a perennial best seller. According to Bertic, Jumanji, based on the book and movie of the same name, continued to do well throughout the year.

So to better gauge the toy market in upcoming years, check out the hot movies and TV shows and stick with some of the classic toy lines. Times may change, and toys do too, but the media and nostalgia are powerful indicators of what may become collectible each year.

Auction Action

By Mike Jacquart

More and more collectors are finding it easier to find their dream toys due to the increased popularity of phone, mail and fax auctions, and last year was no exception.

Several records were set, and many other high prices were realized, but while high-end items hog the spotlight, numerous bargains for hard-to-find toys can also be found at auctions.

It was clear in 1996 that Mickey Mouse still reigns as king at Disney (sorry about that, Mufasa.) Coming off the heels of the rare Mickey and Minnie on a Motorcycle that sold for $30,800 in a 1995 James Julia auction, another Mickey piece — a 1931 tin mechanical bank in the scarce "hands apart" version — sold for a record-setting $36,850 at a Julia auction in Byfield, Mass.

Other Disney Items

The Mickey bank wasn't the only Disney collectible that sold for good prices at auction. Other leading Disney sellers at Julia's auction included an early wind-up celluloid Mickey on a Rocking Horse that sold for $1,650 and a pair of Mickey and Minnie Jam Pots with original lids — a nearly impossible find! — that brought $1,925. The sale offered plenty of other opportunities — many of them more affordable — for Disneyana fans. A 7-inch Mickey celluloid whirligig in working condition went for $550, and a Mickey Mouse Acrobat by Gym Toys, also in working condition, sold for $385. A 6-inch Schuco Donald Duck toy sold for $550 in its original box.

An Ingersoll Mickey Mouse watch was one of the top-selling items at a Manion's auction. The watch, in its original box, sold for $1,110.

A Mickey Mouse Circus Train Set sold for $700 at a Lloyd Ralston auction.

Disney's star mouse was also popular at a Christie's East auction in New York. A charming set of eight, Snow White and the Seven Dwarfs glazed bisque egg cups, sold for $1,294. The colorful set was made in Japan in 1937. German Disney pieces also commanded the attention of collectors. A Goebel Bambi and seated rabbit figure scene from the 1950s sold for $115. A Mickey Mouse doll, with pie-cut eyes, stole a collector's heart for a mere $173. The 12-inch doll by Knickerbocker was made in the 1930s. A Mickey and Minnie Mouse tin litho handcar by Lionel sold for $345.

Dolls

While French bisque dolls made by Emile Jumeau commanded many of the top prices at last year's auctions, a 1950s Madame Alexander doll, the

TOP: This Kewpie charmer sold for $6,400 at a McMasters' doll auction.
ABOVE: This delightful selection of vintage toys was sold at a recent McMasters' auction.
RIGHT: This 1957 Madame Alexander Infant of Prague doll sold for an amazing $51,000 at Theriault's.

prized Infant of Prague, sold for an overwhelming $51,000 at a Theriault's auction of Madame Alexander dolls. The eight-inch Infant of Prague included three religious outfits, two crowns and related accessories.

A 21-inch 1951 McGuffey Ana — a charming hard plastic doll dressed in a pink organdy dress — sold for $17,000 at the same Theriault's sale. Also topping $10,000 was a 21-inch Deborah ballerina Portrait doll that commanded $13,000. The doll, also produced in 1951, sold for around $100 originally.

A 23-inch French bisque Jumeau character doll sold for 10 times its estimate at a Christie's East auction in New York. The doll, dressed in a clown outfit, sold for $36,800. Selling for $4,600 was a French bisque head Steiner. The jointed, 16-inch doll wore a cream silk dress. Commanding $6,900 (slightly under its presale estimate) was a 13-inch Bru Jne Bebe. A 27-inch Jumeau bisque head doll sold for $5,175. In addition to porcelain dolls, other character dolls also sold well — including a pair of 18-inch Ideal Shirley Temple dolls with original buttons and boxes which sold for $3,910.

One doll attracted people from as far away as Europe for a chance to see it in person. The doll — Simon and Halbig's bisque character doll of Mary Pickford — sold for $34,000 at a Theriault's auction in Los Angeles. That was well above the $15,000-$25,000 presale estimate.

Two Jumeau bisque dolls were the top-selling dolls at a Skinner toys and dolls auction. A bisque with a felt hat and French kid boots sold for $9,200 — its presale estimate was $1,000-$2,000. A 19th-century bisque with a satin outfit and straw hat with a presale estimate of $1,500-$2,000 sold for $7,475.

A French hand-pressed bisque "E.J." Jumeau bebe with black complexion doubled its presale by selling for $31,000 at a Theriault's auction in Newport Beach, Calif. Another Jumeau doll of note at the auction was a 19-inch bisque bebe

Buying Dolls at Auction?

Check Out These Tips from the Pros

Pictured is a group of collectible dolls waiting to be sold at a Theriault's auction. The Maryland auction company sells many types of dolls.

By Beth Wheeler

So, you've decided to begin a doll collection and just purchased a fantastic Madame Alexander doll at an unbelievable price at your first auction.

Congratulations! Everyone loves a bargain. But then, you notice the repair marks. The auctioneer didn't mention that, and you didn't attend the preview. A call to the auction house the next day confirms your fears — all purchases are "as-is," and no refund or adjustment will be made.

Unfortunately, this scenario is not uncommon. How can beginners, or seasoned collectors, feel confident about an auction purchase?

Stuart Holbrook, president of Theriault's, a Maryland-based auction firm specializing exclusively in the auction and appraisal of rare, antique and collectible dolls, shares some tips. Actually, he offers a brief seminar for the uninitiated at some of the 10 catalogued auctions and 50 non-catalogued auctions conducted nationwide by the firm each year.

Here are some of Holbrook's tips of the trade.

Know the Auction House

First and foremost, know the auctioneer. Visit an auction held by the company and talk to the bidders. Are bidders pleased with the policies and practices? Does the firm have a reputation for fair dealings?

If you are traveling to an unknown community for a special auction, it may be impractical to attend an additional auction. In that case, contact the Better Business Bureau, doll clubs in the area, a museum, doll shop or Chamber of Commerce. Someone should be able to provide information regarding the reputation of the auction company.

Teddy bears are one of today's hottest collectibles. This bear sold at a Theriault's auction for $2,600, nearly twice the presale estimate.

Setting Your Limits

A large part of attending an auction is knowing why you're there. What item or items do you expect to bid on?

Continued on page 17

Tete that sold for $13,500, twice what the doll would normally be worth. The reason for the added interest was that this doll, like many in the auction, had a provenance or personal story.

Not all of the top-selling French dolls auctioned in 1996 were made by Jumeau. A rare bisque bebe by Huret with a gutta percha body sold for $62,000 at a California Theriault's auction. A French bisque bebe Steiner with a keywind mechanism in its original box and an Au Nain Bleu label nearly tripled its presale by selling for $19,500. Leading Kestner dolls was a pair of German bisque characters known as Max and Moritz. The pair sold for $40,000, nearly four times the estimate.

Future trends of the doll market were also unveiled in a Theriault's auction held in West Palm Beach, Fla. Doll collectors were treated to a large selection of cloth dolls by the Italian firm of Lenci. An unusual Lenci Aladdin doll seated on a cushion sold for $7,750, twice the presale estimate, and a cloth Japanese lady sold for $5,800.

The Florida auction also featured an abundant supply of antique German and French bisque dolls. Tops among the German dolls was a rare bisque character lady by Simon and Halbig, which sold for $24,000 on a presale of $15,000-$20,000. Leading the sale of French dolls was an early Jumeau bisque that brought $10,250.

American dolls and doll houses brought good prices at a Theriault's auction in Seattle. A wood and paper lithographed Bliss doll house sold for $2,400 while a German lithographed doll room with original wood furnishings sold for $2,300 — both well over presale estimates. A pair of Schoenhut wood children — a boy and a girl — realized $2,200.

Kestner, Jumeau, Schoenhut and Kewpie all brought good prices at a McMasters doll auction in Cambridge, Ohio. Highlights of the auction included a 10-inch Kestner bisque head Kewpie — a charmer that sold for $6,400, well over its $3,000-$4,000 estimate. A 17-inch Volland Raggedy Andy sold for an astounding $1,700. A rare 13-inch Schoenhut Bye-Lo exceeded its presale by selling for $2,400.

Barbies

Barbie's popularity was in evidence at a McMasters Barbie auction in Columbus, Ohio. A Ponytail Barbie #3, brunette was the top seller at $1,900 while an American Girl Barbie with light blonde hair sold for $1,425. Over 1,000 Barbies were sold at the two-day sale.

A Barbie doll and accessory exceeded presale estimates at a Just Kids Nostalgia auction. A 1963 vanity set sold for $150, more than double its presale, and a 1965 American Girl Barbie realized $427.

Teddy Bears

A Steiff bear sold for over four times its presale estimate at a Theriault's auction in Seattle. The mohair plush teddy bear sold for $10,800.

Several Steiff bears also exceeded presale estimates at a Theriault's auction in Newport Beach, Calif. A 13-inch plush teddy bear with the trademark button in ear sold for $1,700.

A 13-inch white Steiff brought $4,830 at a Christie's East auction in New York.

Steiff bears aren't the only teddy bears that brought good auction prices last year. A mechanical teddy bear that performed circus-like balancing acts sold for $2,500 at the California Theriault's auction, over four times the estimate. A somersaulting bear sold for $4,100, over three times the estimate.

Vintage Transportation Toys

Whether prewar or postwar, plane, train or automobile, vintage and recent transportation toys sold well at

TOP: This 1936 Dinky car transporter was the top seller ($1,946) at a Christie's London auction.
ABOVE: A four-piece German tin train sold for $22,000 at an Alderfer auction.

auction last year.

Several train sets were featured at a Christie's East auction in New York. A 1920s Ives set with tin Grand Central Station sold for $12,650. Several lots containing Lionel cars and locomotives topped $1,000.

A James Julia auction in Byfield, Mass., featured trains from the Rick Ralston Collection, a 25-year collection that included examples from every major train manufacturer, including Lionel, Bing, American Flyer, Bassett Lowke, Marklin and Ives. Leading the selection was a rare Marklin O Gauge New York Central Hudson locomotive and tender, including Pullman coaches and observation car, that sold for $28,050, far exceeding its $15,000-$20,000 presale estimate. Other noteworthy pieces from the 800-lot collection included a selection of American Flyer trains, including a special Burlington Zephyr three-piece aluminum set engraved for Ralph Budd, president of the Chicago Burlington & Quincy Railroad. The rare set easily passed its $200-$600 estimate to sell for $2,475. A Lionel Blue Comet doubled its estimate to sell for $2,145.

Another train that sold for an eye-opening amount was a four-piece German tin train that brought $22,000 at an Alderfer auction. The train had two passenger cars, a coal car and an engine.

Trains bring top prices at Lloyd Ralston auctions, and nearly 30 trains sold for over $1,000 each at one of the gallery's 1996 auctions. Leading the way were two Lionel sets that each sold for $4,000. One was a Lionel 1587S, and the other was a #60 Adamas Locomotive and Great Western Tender.

A Bing Steam Engine Train Set named The Rhinelander sold for $3,630 at a Bill Bertoia auction in Philadelphia. Each car contained tiny composition figures seated in their compartments or dining tables. The set was designed to show a microcosm of early life on the rails.

An early airplane pedal car sold for $2,970 at a Bill Bertoia auction in Philadelphia. Cast-iron transportation

Buying Dolls at Auction *from page 15*

This one-of-a-kind Mary Pickford doll was commissioned by the actress herself. Only the prototype was made. It sold for $34,000 at a 1996 Theriault's auction.

What do you expect to take home? What are you willing to pay?

1. Attend the preview and select a doll or dolls. Examine the doll carefully, make notes about condition, markings, costume, hair and other details important to you. Check for defects and ask questions. Not every auction company will be as knowledgeable as others about every doll. Does the firm have an expert on hand to provide information about the dolls?

2. Before the auction, set a limit on how much you are willing to spend on each doll. Allow a little flexibility, but don't go beyond the limit. It's very easy to get caught up in the excitement of bidding and spend much more than you had intended, and you may end up hating the doll. Consult a price guide to help establish your bidding limit.

3. Never go to an auction with a friend. Or if you do, don't sit together! Agree ahead of time to bid against one another with no hard feelings later. Someone is going to bid against you, friend or not!

4. Be aware of the tricks experienced auction-goers can use to intimidate the uninitiated. The power bid is one such tactic. As the bidding drives the price ever higher, the power bidder will raise her (or his) hand and/or bidding paddle and keep it up. This gives the impression that the bidder is determined to have the doll at any price. In a successful power play, the uninitiated will be intimidated and drop out of the bidding, allowing the power bidder to win the doll at a lower price.

This rare 26-inch cloth doll by Martha Chase was made around 1900. It doll represents "Mammy," a character from the Uncle Remus stories by Joel Chandler Harris.

Auction Amenities

Comfort during in-person auctions is a priority. Here are some questions you may want answers to before you attend.

Is ample seating available? Are restrooms conveniently located? Are the sound system, ventilation and lighting adequate? Is there a ban on smoking? Are concessions available?

Is the arena arranged to provide an unobstructed view of the dolls being auctioned? Is it accessible to those in wheelchairs? Is there an adequate number of runners available to parade the dolls through the aisles? Are there friendly people assigning numbers and handling accounting?

These guidelines need to be adjusted slightly for on-site estate sales.

Continued on page 19

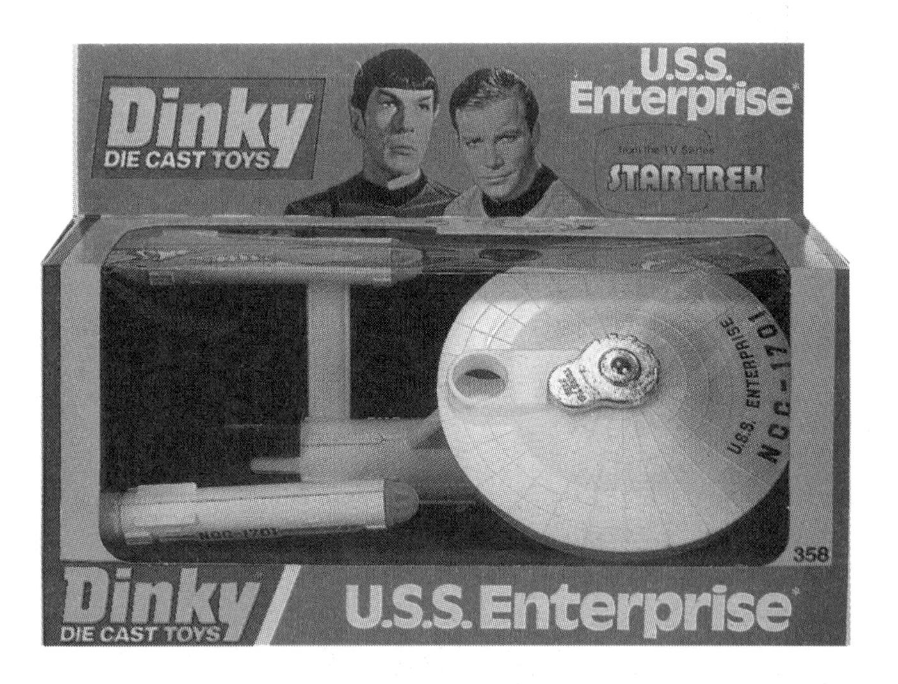

ABOVE: This U.S.S. Enterprise by Dinky Toys commanded the celestial price of $283 at a Christie's South Kensington auction.
BELOW: This arrangement of vintage toys was highlighted at a Noel Barrett auction in 1996.

toys in demand included a Sight Seeing Auto that sold for $2,530 and a Hubley racing boat that realized $6,270. Interest in early fire engine toys also remained high among cast-iron collectors. Two early fire patrol wagons topped $2,500 each.

It's a safe bet the top bidder at a Copake bicycle auction didn't ride his prize home. A hefty $16,225 was offered for an 1889 diamond frame Victor man's hard tire safety bike. One of the oldest bikes in the nation, a Hobby Horse (1819) sold for $8,250. Another highlight was a Shelby Donald Duck 24-inch girl's bicycle from the 1940s that realized $1,760.

Postwar military vehicles, airplanes, cars, trucks and even spaceships — all bearing the Dinky name — were offered at Christie's South Kensington auction of Dinky toys in London, England. A 1936 car transporter hauled in the top bid, realizing $1,946 — more than double its presale estimate. A lot including Lady Penelope's pink Rolls Royce Fab 1 and two Thunderbirds sold for $707. A Lockheed airplane sold for $212, and emergency vehicles included a set of 10 Criterion, Cadillac, Range Rover and Ford ambulances in original boxes ($460).

The Christie's auction also proved that science fiction toys are as popular overseas as they are in the states. The *Star Trek* U.S.S. Enterprise and Klingon Battle Cruiser fell within presale estimates, selling for $238 each. A UFO Interceptor realized $264, and a *Space: 1999* Eagle Transporter exceeded its presale estimate by selling for $354.

While toys from 1960s and 1970s science fiction series remained popular last year, vintage space toys weren't forgotten. Robots sold well at a Julia auction in Byfield, Mass. A Kitahara Lavender Robot by Modern Toys of Japan wasn't even in working condition, but still managed to bring $2,200. A battery-operated moon car by T.N. of Japan, in working condition with its original box, brought the same amount.

A nice selection of transportation

toys were also sold at the Julia auction. A rare 13-inch cast-iron Arcade Mack bus realized a whopping $24,200 — just under the high estimate. Close behind was a cast-iron White Delivery Van by Arcade that sold for $19,525. A Metalcraft Coca-Cola Delivery Truck with electric lights doubled its estimate to bring $2,200.

German airship toys commanded a strong presence at a Noel Barrett auction in New Hope, Pa. A remarkably well-preserved Zeppelin and Airplane Roundego sold for $9,900 — more than tripling its presale estimate. Another airship that soared in value was an early, unmarked bi-wing airplane that sold for $2,860

Planes weren't the only vehicles that brought big bucks at the Barrett sale. An eight-inch clockwork Sidewheeler Boat by Carette sold for $1,650, and a 17-inch Torpedo Boat by Bing realized $1,265. A German Triumph tin motorcycle that only measured 2-3/4-inches long sold for $1,155.

Airplanes and cars brought consistently high prices at a Bertoia auction in Vineland, N.J. A Hubley Friendship Seaplane sold for $5,170, and a 1930s J.E.P. Seaplane with a pontoon went for $4,950. Several German toy cars brought exceptional prices, including a 1902 10-inch Bing Vis-A-Vis that went for $7,920. The popular, cast-iron Andy Gump car realized $6,160.

A Buddy L Trench Digger pressed steel truck from the late 1920s to early 1930s, sold for $4,312 at a Skinner toys and dolls auction. Another top seller was a 40-inch, 1912 German Bing ocean liner that realized $4,025, far exceeding its presale estimate of $1,000-$1,500. A cast-iron train from the 1880s with a locomotive, tender, two gondola

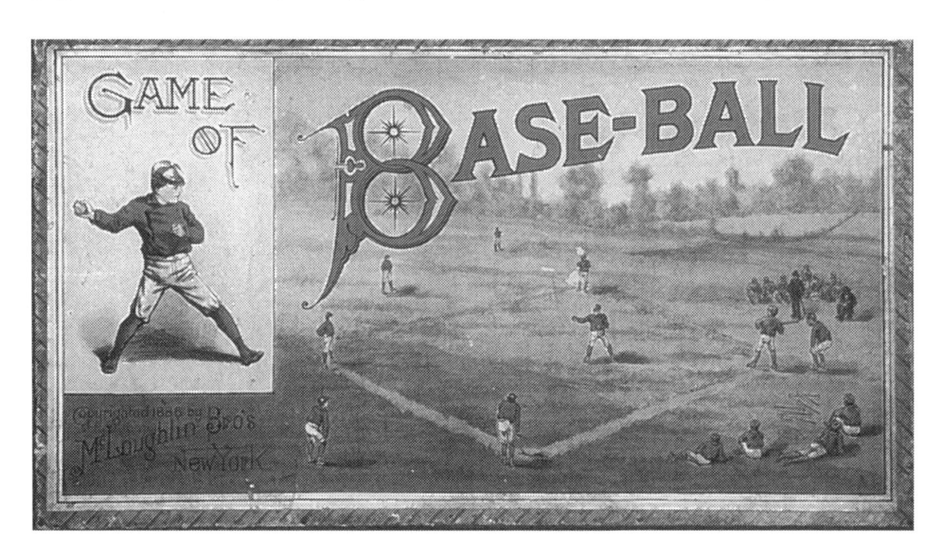

Prewar board games draw much attention at auctions, particularly because of their nostalgia and display value. At a Skinner auction, this 1886 McLoughlin Game of Base-Ball sold for $1,610.

***Buying Dolls at Auction** from page 17*

Auction Rapport

As in a beauty pageant, auctioneers receive ratings from the buyers for poise, grace, good-naturedness, voice quality, voice volume and attentiveness.

How are questions, discrepancies and misunderstandings handled? Discreetly and reasonably?

Is the doll identified well and accurately? Is pertinent background information given regarding the doll?

Does the auction progress smoothly and efficiently, or does the pace lag and the crowd thin with impatience?

Asking the right questions ahead of time and being prepared can help make your doll auction experience more pleasant and, hopefully, more successful.

Beth Wheeler is a free-lance writer from Virginia.

All photos courtesy of Theriault's.

Mattel's Ponytail Barbie #1 from 1959 is an auction favorite, especially in her box. Boxed versions are rare, but with her box, the doll has been known to sell for over $7,000!

In Search of The Perfect Doll

While many nationwide auction houses conduct general auctions, several specialize in dolls of many eras. When looking for that perfect doll to begin or add to your collection, you may want to contact these specialty companies.

Theriault's
P.O. Box 151
Annapolis, MD 21404
800-638-9422 / 410-224-3655

McMasters Doll Auctions
P.O. Box 1755
Cambridge, OH 43725
800-842-3526 / 614-432-4419

Withington, Inc.
RD2 Box 440
Hillsboro, NJ 03244
603-464-3232

Frasher's Doll Auction
Rte. 1 Box 142
Oak Grove, MO 64075
816-625-3786

cars, and original wood box sold for $6,325, nearly double its presale.

Buddy L vehicles also fared well at a Julia sale in Byfield, Mass. A 28-inch Transportation Bus in light green sold for $4,785, and an all-black Buddy L Huckster Truck realized $4,400. A black, red and green Express Line Truck topped its estimate to sell for $2,805, and an all-red International Harvester Sales & Service Truck brought $2,530.

A tin wind-up policeman motorcycle sold for $350 at a House in the Woods auction in Eagle, Wis. An Amos 'n Andy taxi from the 1930s realized $675.

Clowns and Other Toys

Character toys and clowns were also in demand at auctions last year. Schuco's keywind Charlie Chaplin sold for $863, and a flexible 3-1/4-inch Felix the Cat figure sold for $345, both at a Christie's East auction in New York.

Other items at the Christie's auction included a Wolverine tinplate keywind Zilotone with Clown that sold for $2,990, and a pair of tinplate German toys featuring a Ferris wheel that sold for $575.

Dozens of clown toys generated a lot of attention at a Noel Barrett auction in New Hope, Pa. One of the attractions was a 15-inch tin clockwork toy called Two Clowns with Flying Boats that sold for $4,400. Two musician clowns sold for over $3,000 each. An 11-inch Clown Viola Player with composition head sold for $3,850. A selling price of $3,410 was realized for a Clown with Violin, a 13-inch bisque head doll that operated on a clockwork mechanism. Another charmer that brought $2,200 even with some repainting was a clown shown walking wheelbarrow-style while another clown holds his feet.

An always popular classic toy — Schoenhut's Humpty Dumpty Circus — was a feature of a Christie's East auction in New York. The large set, including tent, animals and instruction sheet, sold for $2,070. Various wood animals from the set sold individually for prices ranging from $115 for a group of donkeys to $1,093 for a single antelope.

Popeye, that spinach-munching sailor, fared well at a James Julia sale. A scarce Popeye Overhead Puncher sold within its $2,475 estimate. A 1934 Marx version of the popular Popeye and Olive Oyl on the Roof sold for $1,925. A tin lithographed Popeye the Champ brought $880 and an Australian licorice tin with an image of Popeye with a ship's wheel realized $357. A Felix the Cat hood ornament on a Spanish race car realized $6,600 at a Julia auction.

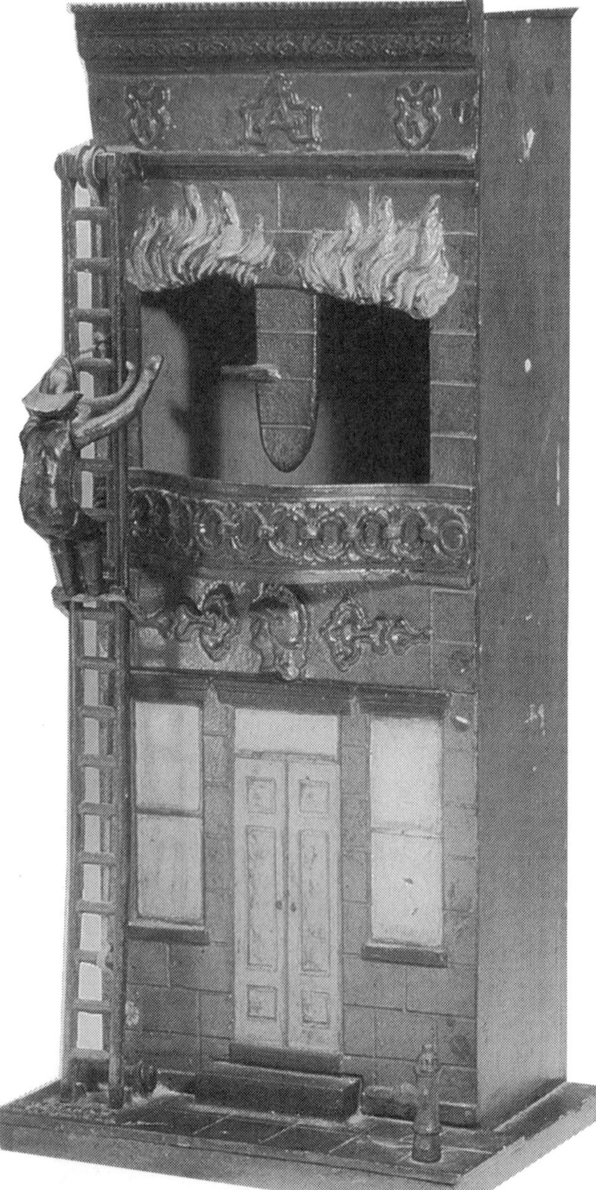

This 1892 Carpenter "burning building" was the top seller ($29,900) at a Skinner auction.

The world's strongest sailor was also a top seller at a Bill Bertoia auction in Philadelphia. Highlights included Popeye and Olive Oyl Jiggers which sold for $3,410, Popeye Express ($2,640), Popeye the Champ ($3,410) and a Popeye Eccentric Airplane ($2,860).

Collectibles from Hanna-Barbera's *Flintstones* and *Ruff and Reddy* cartoon shows topped a Toy Scouts auction. Far exceeding presale estimates was a 1962 boxed pair of *Flintstones* king-sized push puppets that sold for $2,578. The presale was $200-$400. A rare *Ruff and Reddy* squeeze figure sold for $726.

The market for 1960s spy collectibles was strong at Toy Scouts' first TV and movie memorabilia auction. A prototype *Man from U.N.C.L.E.* movie projector gun never released by Marx sold for $1,560, a *Man from U.N.C.L.E.* secret service gun set brought $646, and a James Bond 007 Shooting Attache Case sold for $1,162. Other items that far exceeded presale estimates were a 1968 Winky Dink Drawing Set ($234), Davy Jones Monkees Ring ($72), *Voyage to the Bottom of the Sea* puzzle set ($447), *Gilligan's Island* coloring book ($132), *Waltons* play set ($150) and a *Twelve O'Clock High* card game ($100).

A 1930s Don Winslow Undercover Deputy Torpedo Decoder commanded the top price ($2,106) at a Just Kids Nostalgia auction. Western items were among the other top sellers, including a 1950s Hopalong Cassidy Grape Nuts Flakes cereal box ($1,164), 1950s Roy Rogers gun and holster set ($914) and the first Lone Ranger comic book, issued in 1938 ($1,149).

Board Games

Many board games are not valuable, but that wasn't the case as over 200 late 19th- and early 20th-century board games from the Frank Snook collection were sold at a Noel Barrett auction in New Hope, Pa.

The top seller was J.W. Spear's The Aerial Contest — The Up-to-Date Game. It sold for $2,860, doubling its presale estimate. Highly-detailed lead airplanes added to the game's desirability.

McLoughlin Bros. games, known for their stunning box lithography, were among the other highlights of the sale. A Combination Tiddledy Winks game sold for $742.50, a bicycle race game realized $495 despite not having any playing pieces, and a Game of Tobogganing at Christmas realized $1,650.

Milton Bradley's classic 1911 Checkered Game of Life sold for $220.

Banks

The Hal Kuehl collection took center stage in a Henry/Peirce auction in Homewood, Ill. The collection consisted of vintage and modern mechanical banks as well as cast-iron, glass, ceramic and tin still banks.

Several banks at the nearly 800-lot auction sold for over $5,000 while others sold for prices in the $3,000-$4,000 range. Most top sellers were mechanical banks made in the late 1800s. Commanding top dollar was Paddy and the Pig, an 1882 mechanical from J & E Stevens. It sold for $5,100. Close behind was Clown on Globe, another Stevens mechanical ($5,000).

Still and mechanical banks, most made of cast-iron, also sold well at a Henry/Peirce auction in late 1995. The top-selling bank was the Kiltie, a cast-iron mechanical from J. Harper & Co. (1931). It sold for $1,950. Another best-seller was the popular Darktown Battery by J. & E. Stevens, selling for $1,300.

Vintage banks were also on display at Christie's East auction in New York. The often seen, but still desirable, Humpty Dumpty cast-iron mechanical bank by Shepard Hardware sold for $1,093, as did the familiar Trick Pony mechanical, also by Shepard. Another Shepard bank — an Uncle Sam mechanical — topped presale estimates to sell for $3,450, and a Circus mechanical sold for $7,475. Kyser & Rex's Mammy and Child mechanical bank sold for $6,325, and another popular Black memorabilia piece, the Darktown Battery bank by J & E Stevens, sold for $3,220. One of the unusual banks sold was a Swiss keywind wood musical mechanical bank. The carved clock tower bank from the 1880s made by B.A. Bremond sold for $5,750.

Vintage banks also sold well at a Bertoia auction in Philadelphia. The highlight of the sale was the Calamity Bank, a rare three-figure piece that brought a staggering $44,000. Other top-selling mechanical banks included an unusual Ferris Wheel Bank that brought $21,450, a Horse Race Bank that sold for $24,200 and a Professor Pug Frog Bank that sold for $15,400. The still banks included a rare Canadian Beaver Bank that sold for $1,265 and a Merry Go Round Bank in superb condition that realized $1,595.

Bertoia auctions are known for their mechanical banks, and at a Bertoia sale in Vineland, N.J., the top seller was a Bread Winners Bank that realized $19,800.

A rare cast-iron Carpenter burning building made in 1892 sold for a staggering $29,900 at a Skinner auction. The item nearly doubled its $15,000 estimate despite missing a woman figure in the window. The piece featured a climbing fireman on a ladder.

Cast-iron banks from the 1930s and earlier can bring top auction prices. This Kiltie bank sold at auction for $1,950.

Guide to Major Auction Houses

Want to find out more about some of the nation's leading auction houses or dates of 1997 auctions? Here are some addresses:

- ✔ **Bill Bertoia Auctions**, 2413 Madison Ave., Vineland, NJ 08360 (609-692-1881).
- ✔ **Christie's**, 502 Park Ave., New York, NY 10022 (212-548-1119)
- ✔ **Henry/Peirce Auctioneers**, 1525 S. Arcadian Dr., New Berlin, WI 53151 (414-797-7933)
- ✔ **James Julia Inc.**, P.O. Box 830, Fairfield, ME 04937 (207-453-7125)
- ✔ **Just Kids Nostalgia**, 310 New York Ave., Huntington, NY 11743 (516-423-8449)
- ✔ **Lloyd Ralston Toys**, 109 Glover Ave., Norwalk, CT 06850 (203-845-0033)
- ✔ **Noel Barrett Antiques and Auctions**, P.O. Box 300, Carversville, PA 18913 (215-297-5109)
- ✔ **Skinner Inc.**, 357 Main St., Bolton, MA 01740 (508-779-6241)
- ✔ **Sotheby's**, 1334 York Ave., New York, NY 10021 (212-606-7176)
- ✔ **Toy Scouts**, 137 Casterton Ave., Akron, OH 44303 (330-836-0668)

Avoid Auction Pitfalls By Learning the Rules of the Game

By Mike Jacquart

Caveat emptor — Latin for "Let the buyer beware."

That phrase offers good advice for collectors who've never bought items at auction, several toy dealers said.

That's because while dealers believe most of the people in their field are honest, unscrupulous people exist in any type of business, they said.

"Whether it's an auction over the telephone or in person, when there's money to be made, you're going to have some shady people," said Bill Bruegman of Toy Scouts, an Ohio auction company.

How do bidders avoid dishonest dealers? Bruegman recommends bidders read the rules and become familiar with dealers.

"Ask questions and check out people's reputations," he advises. "Do they stand behind their guarantees?"

Inexperience can Mean Paying Too Much

But dealers believe the biggest problem novice bidders face is paying too much for items due to their lack of experience. That's why dealers suggest novices learn as much as possible prior to auctions.

"Get there early," suggests Bob Peirce, who holds auctions of still and mechanical banks in the Midwest. "Ask questions like, 'How does this bank compare to other banks?' or 'Is this paint in Good condition?'

"Look at items carefully, and set a limit of what you're going to pay," Peirce added. "The more knowledge you can acquire, the better off you're going to be."

Presale estimates — price ranges of what items are expected to sell for — aren't mandatory but they can be helpful, Bruegman said.

"I didn't do this when I started [holding auctions] but people like having an idea what to bid," he said. "It gives them a starting point so they're not starting too high or too low."

Dealers Face Risks Too

But bidders aren't the only people who face risks at auctions. Dealers must be wary of people who don't pay — although many said that rarely happens because bidders are usually required to sign contracts called bid sheets that list the amount of the bid. In addition, because many collectors are willing to spend $5-$10 for auction catalogs, dealers believe they are serious bidders.

"If you give away a kitten," Bruegman theorized, "you don't know how that person will treat it — a lot of people will take a free kitten. But if you're willing to pay $10 for that kitten, chances are you're a good pet owner."

Some dealers will accept partial payments, but that's not common, Peirce said. In addition, dealers must also take into account the month or two it can take for a check to clear from an absentee bidder.

Some dealers charge buyer's premiums — usually 10-15 percent tacked onto the bid — to help offset the costs of postage, advertisements, labor, and rental space.

Bruegman said he doesn't include buyer's premiums in the items he auctions. "We don't have the overhead of a large auction house," he said. "I usually cover my costs with the catalogs I sell."

Peirce charges 10 percent premiums at his auctions. "It helps offset our costs a lot," he said.

Phone Auctions are Growing Trend

Auctions are generally divided into live or "on premise" auctions, in which bids are made in person, and open or telephone auctions, in which bidders phone or fax in bids.

"As busy as people are today, phone auctions seem to be the craze of the 1990s," Bruegman said. "People can bid from the comfort of their home. The flip side of the coin is that it's harder to know what you're getting. Photos and descriptions of items are very important in this kind of auction."

That's why Peirce recommends only experienced collectors participate in these types of auctions.

Live Auctions are Best for Novices

"Going to the auction is the best way to see what's there," Peirce said. "Those starting out can see what they're getting, they can pick up merchandise, and they can meet other collectors and see what kind of interest is out there for specific pieces."

While high-end items tend to get more publicity, auctions also offer better values than many collectors realize, dealers said. Bruegman estimates 30 to 50 percent of items at auction sell for prices below market value.

"You find not only rare items at auctions, but there are some good bargains as well," he said.

"While it's true auction prices can be misleading because some people will bid a lot for an item they really want," Peirce said, "auctions, generally speaking, serve as a good frame of reference for people out of touch with current prices."

So now you're ready to cast or phone in your first bid. What last minute advice can help?

"Look for quality, not just the lowest price. If you're only worried about bidding low, you might not be getting the quality item you want," Peirce said.

"Ask *any* unanswered questions you might have," Bruegman said. "That's especially important in phone auctions. Is there a guarantee? If you're going to go out of town after placing your bid, you should consider making a sealed bid in which you have a range with a maximum limit. Otherwise, you might prefer a call back to see how your bid fares to other bidders.

"The point is," Bruegman added, "know the rules and know your options within the rules."

A Look at What Was, Wasn't and Will Be Hot

By Bill Underwood

It's a big job trying to stay abreast of the fast-paced toy collecting hobby every year. But some one's got to do it, and *Toy Shop* is pleased to be your source of that information.

Every year, the secondary market fluctuates based on many factors — hot movies or TV shows, limited-edition introductions, reissues or just plain whimsy and speculation.

What was hot last year may now have cooled considerably in the fickle collectors' world. Long-stagnant pieces, however, may also be enjoying a resurgence in popularity.

We've eliminated the guesswork for you. We've surveyed top toy dealers, asking what was, wasn't, and will be hot in the world of collectible toys.

Also featured below are lists of the top toys in various collecting categories — from action figures to vehicles. The value listings are culled from a variety of sources — print retail ads, dealer price lists, observed prices at toy shows, auction prices realized and personal input from expert dealers and collectors nationwide.

This market update gives you a sneak peek at information in the 888-page *1997 Toys & Prices*. Produced by the publishers of *Toy Shop*, the updated and expanded book is available for $17.95 plus shipping and handling. To order, call Krause Publications at 800-258-0929.

The Top 10 Action Figures
(excluding G.I. Joe and Captain Action)
(in Mint in Box condition)

1. Batgirl, Comic Heroines, Ideal, 1967 .... $4,500
2. Wonder Woman, Comic Heroines, Ideal, 1967 3,000
3. Mera, Comic Heroines, Ideal, 1967 3,000
4. Supergirl, Comic Heroines, Ideal, 1967 3,000
5. Scorpio, Major Matt Mason, Mattel, 1967-70 1,500
6. Romulan, Star Trek, Mego, 1976 600
7. Mission Team Four-Pack, Major Matt Mason, Mattel, 1966-69 575
8. Bruce Wayne, World's Greatest Super Heroes, boxed, Montgomery Ward exclusive, Mego, 1974 500
9. Jaws, James Bond: Moonraker, Mego, 1979 500
10. Speedy, Teen Titans, Mego, 1976 500

The Top 20 G.I. Joe Figures/Sets
(in MIP condition)

1. Foreign Soldiers of the World, #8111-83, Action Soldiers of the World, 1968 .... $4,250
2. Adventure Pack, Army Bivouac Series, #7549-83, Action Soldier Series, 1968 3,500
3. Dress Parade Adventure Pack, #8009.83, Action Soldier Series, 1968 3,500
4. Talking Adventure Pack, Mountain Troop Series, #7557-83, Action Soldier Series, 1968 .. 3,500
5. Crash Crew Fire Truck Set, #8040, Action Pilot Series, 1967 3,500
6. Jungle Fighter Set, #7732, Action Marine Series, 1967 3,500
7. Negro Adventurer, #7905, Adventures of G.I. Joe, 1969 3,500
8. G.I. Nurse, #8060, Action Girl Series, 1967 3,500
9. Talking Landing Signal Officer Set, #90621, Action Sailor Series, 1968 3,500
10. Talking Shore Patrol Set, #90612, Action Sailor Series, 1968 3,500
11. Military Police Uniform Set, #7539, Action Soldier Series, 1967 3,500
12. Talking Action Pilot, #7890, Action Pilot Series, 1967 3,500
13. Talking Adventure Pack, Special Forces Equip., #90532, Action Soldier Series, 1968 .. 3,500
14. Navy Scuba Set, #7643-83, Action Sailor Series, 1968 3,250
15. Green Beret, #7536, Action Soldier Series, 1966 3,000
16. Talking Adventure Pack w/Field Pack Equip., #90712, Action Marine Series, 1968 .. 3,000
17. Aquanaut, #7910, Adventures of G.I. Joe, 1969 3,000
18. Talking Adventure Pack, Command Post Equip., #90517, Action Soldier Series, 1968 ... 3,000
19. Talking Adventure Pack, Bivouac Equip., #90513, Action Soldier Series, 1968 3,000
20. Talking Adventure Pack and Tent Set, #90711, Action Marine Series, 1968 3,000

Action Figures

From G.I. Joe to Luke Skywalker, action figures are the undisputed heroes of the toy collecting world. And with the seemingly endless supply of new figures flooding the market each year, there are plenty of heroes to choose from.

G.I. Joe seized the high ground in 1996 with the long-awaited release of the G.I. Joe Masterpiece Edition, a flawless reproduction of the original 1964 Joe, complete with a book chronicling his origins by Joe creator Don Levine and author John Michlig.

[The set was pulled from shelves in late 1996 due to a mold problem on the figures.]

Also making an appearance 1996 was Kenner's G.I. Joe Classic Collection. These new Joes featured radically redesigned bodies with more articulation than original 1964 figures. Kenner hopes this new Joe will carry the line into the 21st century. A female helicopter pilot — the first female 12-inch in the line in 30 years — should spark collectors' interest.

The 3-3/4-inch line of G.I. Joes was discontinued in 1994, but a Rhode Island dealer said his hottest selling action figures are the small Joes Mint in Package.

The Star Wars universe continued to expand with Kenner's Power of the Force series of figures and accessories. Dealers quickly hoarded short-packed figures like the first issue Princess Leia, but Kenner continued to release all figures from its first wave (Leia included) — although in smaller quantities. What this will do to prices on the secondary market is speculation.

Also hot in 1996 were Kenner's 12-inch *Star Wars* figures based on characters from the original film.

Look for Trendmasters' well-made and well-marketed *Independence Day* figures and accessories to increase in collectibility and value.

The Top 10 Mechanical Banks
(in Excellent condition)

1. Jonah and the Whale, Stevens, J.& E., 1880s $55,000
2. Mikado, Kyser & Rex, 1886 55,000
3. Circus Bank, Shepard Hardware, 1888 45,000
4. Girl Skipping Rope, Stevens, J.& E., 1890 45,000
5. Calamity, Stevens, J.& E., 1905 35,000
6. Rollerskating Bank, Kyser & Rex, 1880s 30,000
7. Harlequin, Stevens, J.& E., 1907 22,000
8. Motor Bank, Kyser & Rex, 1889 20,000
9. Picture Gallery Bank, Shepard Hardware, 1885 20,000
10. Called Out Bank, Stevens, J.& E., 1900 20,000

The Top 10 Ertl Die-Cast Vehicle Banks 1981-1990
(in Mint condition)

1. Gulf Refining, #9211, 1984 ....$2,500
2. Texaco #1 Sampler, #2128, 1984 1,500
3. Texaco #2 Sampler, #9238, 1985 1,050
4. Texaco #1, #2128, 1984 950
5. Texaco #1, #2128, 1986 950
6. Durona Productions, #1321, 1982 725
7. Harley-Davidson #1, #9784, 1988 700
8. Texaco #3 Sampler, #9396, 1986 675
9. Texaco #2, #9238, 1985 650
10. Harley-Davidson #2, #9135UO, 1989 650

Banks

Cast-iron still and mechanical banks were still the hottest collectibles in this category with J.E. Stevens' Jonah and the Whale topping the list of mechanicals at a whopping $55,000.

In the more affordable range were die-cast banks (many made by the Ertl Company) with top 10 prices ranging anywhere from $650 to $2,500, reflecting little change from last year.

Dealer Don Potts from Fort Worth, Texas said Ertl's Route 66 series has been a hot seller. The Texaco series also looks like a safe investment, Potts added.

Collectors should beware of cast-iron reproductions. One expert on collectible banks said repros can be spotted by poor paint jobs, different gloss and less weight due to inferior grades of iron.

The Top 10 Still Banks
(in Excellent condition)

1. Indiana Paddle Wheeler, Unknown, 1896 ....$8,000
2. Tug Boat, Unknown 7,500
3. San Gabriel Mission, Unknown 7,500
4. Coin Registering Bank, Kyser & Rex, 1890 6,500
5. Electric Railroad, Shimer Toy, 1893 6,000
6. Boston State House, Smith & Egge, 1800s 6,000
7. Hippo, Unknown 6,000
8. Chanticleer (Rooster), Unknown, 1911 6,000
9. Dormer Bank, Unknown 6,000
10. Camera Bank, Wrightsville Hardware, 1800s 5,000

Barbie

Since her coming out party in 1959, Barbie has been Le Grande Dame of not just dolls, but toys in general. Today, you can't go wrong with vintage Barbies and accessories from the '60s, said Barbie expert Renay Williams.

Although the vintage pieces are probably the safest investments, there are plenty of newer Barbies that command high prices on the secondary market. Barbies featuring designer apparel by Bob Mackie were particularly hot. A first issue Gold Bob Mackie can go for upwards of $700 to $900.

The Top 10 Barbie Dolls
(in MIB condition)

1. Ponytail Barbie #1, brunette, 1959 .... $9,500
2. Ponytail Barbie #1, blonde, 1959 7,500
3. Ponytail Barbie #2, brunette, 1959 7,000
4. Ponytail Barbie #2, blonde, 1959 5,000
5. American Girl Side-Part Barbie, 1965 3,500
6. Color Magic Barbie, black hair, 1966 2,500
7. Pink Jubilee Barbie, only 1200 made, 1989 1,900
8. American Girl Barbie, "Color Magic Face," 1965 1,800
9. Color Magic Barbie, blonde, 1966 1,500
10. Harvey Nichols Special Edition (250 made), 1995 1,500

The adult market for Barbie collectibles has not escaped Mattel's attention. The company released a dizzying array of limited-edition figures, as well as store exclusives from FAO Schwarz, Spiegel, J.C. Penney and Sears.

Williams predicted the FAO Schwarz Antique Rose Barbie will have legs investment wise. But with a price tag of $250, the doll will be out of reach for many collectors.

Williams also predicted the new Bob Mackie Moon Goddess figure to do well. And as long as Mattel does not flood the market with 1996 Holiday Barbies, as it did in 1995, those figures will increase in value.

The coming year will also tell the fortune of Mattel's Pink Splendor Barbie. With its winter 1996 release, the doll carried the highest price tag on a Barbie to date — $900. Time will tell if it will become a timeless classic or if Mattel may need to rethink the price ceiling, even for collectors' edition dolls.

The Top 10 Barbie Fashions
(in NRFB condition)

1. Gay Parisienne, #964 .... $2,500
2. Roman Holiday, #968 2,500
3. Easter Parade, #971 2,500
4. Pan American Stewardess, #1678 1,500
5. Here Comes The Groom, #1426 1,000
6. Best Man, #1425 1,000
7. Shimmering Magic, #1664 950
8. Commuter Set, #916 850
9. Gold n' Glamour, #1647 850
10. Beautiful Bride, #1698 850

The Top 10 Battery-Operated Toys
(in Mint condition)

1. Big Loo, Marx, 1960s $2,010
2. Bubble Blowing Popeye, Linemar, 1950s 1,625
3. Mickey the Magician, Linemar, 1960s 1,625
4. Smoking Spaceman, Linemar, 1950s 1,600
5. Drumming Mickey Mouse, Linemar, 1950s 1,425
6. Kooky-Spooky Whistling Tree, Marx, 1950s 1,000
7. Tarzan, Marusan, 1960s 1,000
8. B-58 Hustler Jet, Marx, 1950s 925
9. Disney Fire Engine, Linemar, 1950s 870
10. Television Spaceman, Alps, 1960s 825

Character Toys

According to dealer Jon Thurmond, probably the hottest character toys today are G.I. Joe and *Star Wars* figures with Barbie coming in a close third.

The release or remake of a movie or TV show almost always increases the value of a character collectible. For instance, when the *Casper* movie came out in 1995, *Casper* collectibles were in great demand. They have not, however, continued to climb in value.

The Japanese market for character collectibles has gotten very hot, Thurmond added. Characters from the *Gremlins* movies are highly prized among Japanese collectors, especially the Gizmo character. Also hot in Japan is anything from *Planet of the Apes*.

In the United States, *Six Million Dollar Man* collectibles picked up, as did figures and accessories in Mattel's 1960s Major Matt Mason space line. Looney Tunes characters enjoyed a revival and will probably continue to do so with the winter 1996 release of Warner Brothers' *Space Jam*.

Walt Disney's *Hunchback of Notre Dame* stumbled in 1996 in the wake of the more upbeat and eagerly-anticipated *Toy Story*. With the Halloween, 1996, release of the *Toy Story* video, a renewed interest and new toys are expected.

The Top 25 Character Toys/Collectibles
(in Mint condition)

1. Superman Candy Ring, Leader Candy, 1940..........$20,000
2. Superman-Tim Club Ring, 1940s..........15,000
3. Superman Patch, 1939..........15,000
4. Superman Trading Cards, Gum Inc., 1940..........10,000
5. Popeye the Heavy Hitter, Chein..........6,750
6. Popeye the Acrobat, Marx..........5,500
7. Superman Hood Ornament, Lee, 1940s..........5,000
8. Popeye the Champ, Marx, 1936..........4,600
9. Superman Statue, Syracuse Ornament, 1942..........4,000
10. Captain Marvel Sirocco Figurine, Fawcett, 1940s....3,500
11. Captain Marvel Statuette, Fawcett, 1940s..........3,200
12. Mickey Mouse Radio, Emerson, 1934..........3,000
13. Dick Tracy Lamp, 1950s..........3,000
14. Superman Roll-Over Plane, Marx, 1940..........3,000
15. Popeye Express, Marx, 1936..........2,900
16. Popeye & His Punching Bag Toy, Chein, 1930s......2,500
17. Man from U.N.C.L.E. THRUSH Rifle, Ideal..........2,500
18. Alice in Wonderland Cookie Jar, Regal, ...2,500
19. Superman Tank, Linemar, 1958..........2,500
20. Superman Doll, Ideal, 1940..........2,500
21. Twelve Faces of Jack display, Disney's *Nightmare Before Christmas*..........2,300
22. Superman Action Game, American Toy Works........2,000
23. Popeye Handcar, Marx, 1935..........2,000
24. Mickey Mouse Tricycle Toy, Steiff, 1932..........2,000
25. Green Hornet Dashboard, Remco, 1966..........2,000

The Top 10 Coloring Books
(in Mint condition)

1. Courageous Cat and Minute Mouse, Artcraft..........$175
2. Funny Company, Whitman, 1966..........130
3. George of the Jungle, Whitman, 1967..........130
4. Shirley Temple Crosses the Country, Saalfield..........125
5. John Wayne Coloring Book, Saalfield, 1951..........125
6. Three Stooges, Lowe..........125
7. Touché Turtle, Whitman..........125
8. Charlie Chaplin Up in the Air, M.A. Donahue..........125
9. Walt Disney's Mickey Mouse Coloring Book, Whitman, 1970s..........105
10. Kimba the White Lion, Saalfield..........100

The Top 10 A.C. Gilbert Erector Sets
(in Excellent condition)

1. Set #10, eight-drawer oak chest, 1927..........$7,500
2. Set #10, oak box, 1929-31..........7,500
3. Set #10, wood box, 1920-26..........7,500
4. Set #10, nine-drawer oak chest, 1928..........7,500
5. Set #9, wood box, 1929-32..........4,200
6. Set #8-1/2, wood box, 1931-32..........3,000
7. Set #7-1/2, wood box, 1914..........1,500
8. Set #12-1/2, metal box, 1948-50..........1,300
9. Set #12-1/2, metal box, 1956-57..........1,300
10. Set #10093, metal box, 1959..........1,250

The Top 25 Fisher-Price Toys
(in Mint in Box condition)

1. Skipper Sam, #155,1934..........$2,800
2. Popeye the Sailor, #703, 1936..........2,400
3. Popeye, #700, 1935..........2,200
4. Popeye Cowboy, #705, 1937..........2,100
5. Woodsy-Wee Circus, #201, 1931..........2,100
6. Mickey Mouse Band, #530, 1935..........2,100
7. Dog Cart Donald, #149, 1936..........2,000
8. Drummer Bear, #102, 1931..........2,000
9. Doughboy Donald, #744, 1942..........2,000
10. Chubby Chief, #110, 1932..........2,000
11. Dumbo Circus Racer, #738, 1941..........2,000
12. Doc & Dopey Dwarfs, #770, 1938..........2,000
13. Barky Puppy, #103, 1931..........2,000
14. Doctor Doodle, #100, 1931..........2,000
15. Granny Doodle & Family, #101, 1933..........2,000
16. Granny Doodle, #101, 1931..........2,000
17. Raggedy Ann & Andy, #711, 1941..........1,975
18. Scotty Dog, #710, 1933..........1,900
19. Pushy Doddle, #507, 1933..........1,900
20. Walt Disney's Easter Parade, #475, 1936..........1,900
21. Go'N Back Jumbo, #360, 1931..........1,800
22. Donald Cart, #469, 1940..........1,800
23. Lucky Monk, #109, 1932..........1,800
24. Pushy Piggy, #500, 1932..........1,800
25. Mickey Mouse Choo-Choo, #432, 1938..........1,800

Erector Sets

Original A.C. Gilbert sets commanded the highest premiums in this category, particularly the earliest ones bearing the company's earlier Mysto Manufacturing name.

Set #10, an eight-drawer oak chest from 1927, tops the *Toys & Prices* list at $7,500 Mint in Box.

Although still a relatively small subset in the toy collecting field, prices suggest construction sets may have a promising future for collectors.

Games

According to George Newcomb of Plymouth Rock Toy Company in Mass., TV-related and space games remained as popular as ever.

Although movie-related games often do well, they have not enjoyed the enduring popularity of games centered around long-running TV shows like *I Dream of*

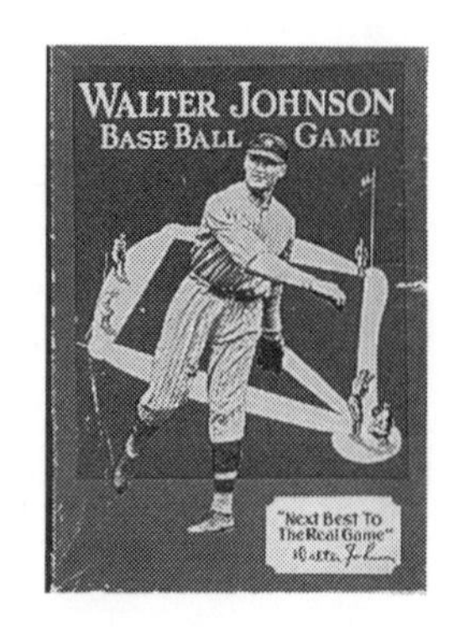

WALTER JOHNSON
BASE BALL GAME
"Next Best To The Real Game"

Jeannie, *The Partridge Family* and *The Howdy Doody Show*.

Sports games continued to do well, especially those featuring a specific sports celebrity. Baseball greats like Willie Mays and Pee Wee Reese have board games bearing their names and photos. Some science fiction games also were safe bets, particularly perennially hot favorites like *Lost in Space* and *Star Wars*.

The Top 10 Prewar Sports Games
(in Excellent condition)

1. New Parlor Game of Baseball, Sumner, 1896....$10,000
2. Champion Baseball Game, The, Schultz, 1889 6,800
3. Zimmer Baseball Game, McLoughlin Bros. 1885 6,000
4. Egerton R. Williams Popular Indoor Baseball Game, Hatch, 1886 5,000
5. Golf, Schoenhut, 1900....................5,000
6. Great Mails Baseball Game, Walter Mails Baseball Game, 1919 4,100
7. American League Fan Craze Card Game, Fan Craze, 1904....................3,250
8. Parlor Base Ball, Game of, McLoughlin, 1892..... 2,900
9. Diamond Game of Baseball, The, McLoughlin Bros., 1894 2,150
10. Bing Miller Base Ball Game, Ryan, 1932 2,000

The Top 10 Postwar Sports Games
(in Mint condition)

1. Red Barber's Big League Baseball Game, G & R Anthony, 1950s....................$900
2. Win A Card Trading Card Game, Milton Bradley, 1965 900
3. Strike Three, Tone Products, 1948....................725
4. Willie Mays "Say Hey," Toy Development, 1954 525
5. Big Six Sports Games, Gardner, 1950s....................450
6. Baseball Card Game, Official, Milton Bradley, 1965....................450
7. Pee Wee Reese Marble Game, Pee Wee Enterprises, 1956 450
8. Willie Mays "Say Hey" Baseball, Centennial Games, 1958 450
9. Willie Mays Push Button Baseball, Eldon, 1965 450
10. Bart Starr Quarterback Game, 1960s 450

The Top 20 Prewar Character Games
(in Excellent condition)

1. Bulls and Bears, McLoughlin Bros., 1896.........$13,000
2. Little Fireman Game, McLoughlin Bros., 1897.....6,000
3. Teddy's Ride from Oyster Bay to Albany, Jesse Crandall, 1899 5,500
4. National Game of the American Eagle, Ives, 1844....................4,000
5. Rival Policemen, McLoughlin Bros., 1896 4,000
6. Darrow Monopoly, Charles Darrow, 1934 4,000
7. Man in the Moon, McLoughlin Bros., 1901 3,500
8. Fire Alarm Game, Parker Brothers, 1899...............2,300
9. Tom Barker Card Game, 1913....................2,300
10. Merry-Go-Round, Chaffee & Selchow, 1898.......2,000
11. Detective, The Game of, Bliss, 1889....................2,000
12. Newsboy, Game of the, Bliss, 1890 2,000
13. Watermelon Patch Game, McLoughlin Bros., 1896....................2,000
14. Hand of Fate, McLoughlin Bros., 1901 2,000
15. Stanley in Africa, Bliss, 1891....................1,800
16. Yale Harvard Game, McLoughlin Bros., 1890.....1,775
17. American Revolution, New Game of The, Lorenzo Burge, 1844....................1,600
18. Shopping, Game of, Bliss, 1891....................1,500
19. Train for Boston, Parker Brothers, 1900 1,500
20. Cake Walk Game, Parker Brothers, 1900s 1,500

The Top 20 Postwar Character Games
(in Mint condition)

1. Elvis Presley Game, Teen Age Games, 1957........$1,000
2. Man from U.N.C.L.E. Target Game, Marx, 1966......550
3. Mickey Mouse Haunted House Bagatelle, 1950s......550
4. Jonny Quest Game, Transogram, 1964....................500
5. Green Hornet Quick Switch Game, Milton Bradley, 1966 475
6. Monster Lab, Ideal, 1964 450
7. Mighty Hercules Game, Hasbro, 1963 450
8. New Avengers Shooting Game, Denys Fisher, 1976 440
9. Godzilla, Ideal, 1960s....................400
10. Batman Batarang Toss, Pressman, 1966....................400
11. Superman, Adventures of, Milton Bradley, 1940s ...375
12. James Bond Message From M Game, Ideal, 1966..350
13. Gilligan's Island, Game Gems, 1965 325
14. Wolfman Mystery Game, Hasbro, 1963 320
15. Thing Ding Robot Game, Schaper, 1961....................310
16. Boom Or Bust, Parker Brothers, 1951 300
17. Jackie Gleason's Story Stage Game, Utopia, 1955..300
18. Creature From The Black Lagoon, Hasbro, 1963....280
19. Outer Limits, Milton Bradley, 1964....................275
20. Justice League of America, Hasbro, 1967 275

Word games, strategy games and generic games like Rook, Pit, Monopoly and Trivial Pursuit traditionally make poor investments, experts said. A prewar Monopoly or Pit game may look nice and vintage, but it won't reap collector prices.

Toy Guns

When it comes to toy guns, anything related to a popular TV show is usually a good investment, said dealer George Newcomb. A Mint Paladin set can command upwards of $1,000, while a Roy Rogers set in similar condition can sell for as high as $3,000.

One of the factors that continues to drive this market is the crossover effect of movie and TV collectors competing for a limited number of collectibles.

As a general rule, Newcomb recommended collectors avoid cast-iron guns in favor of die-cast versions. Mint holster sets from the 1950s and 1960s also are highly prized.

The Top 10 Toy Guns/Sets
(in MIB condition)

1. Man From U.N.C.L.E. THRUSH Rifle, Ideal, 1966 ....$2,500
2. Lost In Space Roto-Jet Gun Set, Mattel, 1966 2,000
3. Man From U.N.C.L.E. Attache Case, Ideal, 1965 1,500
4. Cap Gun Store Display, Nichols, 1950s 950
5. Showdown Set with Three Shootin' Shell Guns, Mattel, 1958 950
6. Man From U.N.C.L.E. Attache Case, Lone Star, 1966........ 850
7. Roy Rogers Double Gun & Holster Set, Classy, 1950s 800
8. Lost In Space Helmet and Gun Set, Remco, 1967 800
9. Man From U.N.C.L.E. Napoleon Solo Gun Set, Ideal, 1965 800
10. Man From U.N.C.L.E. Attache Case, Lone Star, 1966........750

The Top 10 Lunch Boxes
(in Near Mint condition)

1. 240 Robert, Steel, Aladdin, 1978....................$2,500
2. Toppie Elephant, Steel, American Thermos, 1957 1,500
3. Home Town Airport Dome, Steel, King Seeley Thermos, 1960 950
4. Knight in Armor, Steel, Universal, 1959 820
5. Ballerina, Vinyl, Universal, 1960s 800
6. Dudley Do-Right, Steel, Universal, 1962.............800
7. Superman, Steel, Universal, 1954........................800
8. Bullwinkle & Rocky, Steel, Universal, 1962........800
9. Underdog, Steel, Okay Industries, 1974..............800
10. Little Friends, Steel, Aladdin, 1982...................760

Lunch Boxes

Whether metal or plastic, lunch boxes were as popular as ever because of their nostalgia and display value. Larry Aikins of P&L Collectibles in Texas believes types of lunch boxes make good investments.

"Lunch boxes based on TV shows always move great," Aikins said.

The Jetsons, Lost in Space, and *The Brady Bunch* were three desirable boxes last year, Aikins added.

Just the Thermos bottle alone that came with some lunch boxes can command upwards of $120.

As for future investments, Aikins believes soft, pliable lunch boxes are becoming more popular with

retail consumers, which could result in hard plastic boxes being phased out of the market. As a result, plastics will be very collectible, Aikins said.

Marx Play Sets

Marx play sets were among the more stable toy collecting investments one can make, said longtime California dealer Glenn Holcomb.

Prices on these sets tend to remain stable from year to year. In 1996, the Ben Hur, Gunsmoke and Johnny Tremain Revolutionary War sets were among the top collectibles in this category with Johnny Ringo topping the list at $2,500. Another favorite, The Untouchables, was introduced in 1961 and is considered a bargain today at $1,500.

The Top 25 Marx Play Sets
(in MIB condition)

1. Johnny Ringo Western Frontier Set, #4784, 1959,	$2,500
2. Johnny Tremain Revolutionary War, #3402, 1957	2,000
3. Civil War Centennial, #5929, 1961	2,000
4. Gunsmoke Dodge City, #4268, 1960	2,000
5. Fire House, #4820	2,000
6. Ben Hur, #4701	1,800
7. Sears Store, #5490, 1961	1,800
8. Custer's Last Stand, #4670, 1963	1,800
9. Wagon Train, #4888	1,500
10. Untouchables, #4676, 1961	1,500
11. World War II European Theatre, #5949	1,500
12. Sword in the Stone, British only Disney release	1,500
13. Jungle Jim, #3706, 1957	1,400
14. Walt Disney's Zorro, #3754, 1958	1,300
15. Ben Hur, #4702, 1959	1,250
16. Adventures of Robin Hood, #4722, 1956	1,250
17. Roy Rogers Rodeo Ranch, #3986R, 1958	1,250
18. Rin Tin Tin at Fort Apache, #3686R, 1956	1,200
19. Walt Disney's Zorro, #3758, 1958	1,200
20. Skyscraper, #5449	1,200
21. Skyscraper, #5450	1,200
22. Battle of the Blue & Gray, #4744, 1963	1,200
23. Battle of the Blue & Gray, #4658	1,200
24. I.G.Y. Arctic Satellite Base, #4800, 1959	1,175
25. Walt Disney's Zorro, #3753, 1958	1,150

This category of collectibles is so stable, Holcomb said, the only thing that could hurt the market would be a factory find of some of the rarer sets. And factory finds are becoming rarer with each passing year.

Model Kits

Not much has changed in this category since last year. Cult favorites from the 1960s like Aurora's classic Godzilla's Go-Cart were still among the most sought-after figural model kits, especially if sealed and in the original boxes.

Figure models inspired by 1960s TV shows remained popular with collectors, including *Lost In Space*, *The Munsters* and *The Addams Family.* Any Aurora figure kits from the 1960s are sure to be winners as future investments.

The Top 10 Figural Model Kits
(in Mint In Box condition)

1. Godzilla's Go-Cart, Aurora, 1966	$3,000
2. Lost In Space, large kit w/ chariot, Aurora, 1966	1,300
3. Munsters, (Living Room) Aurora, 1964	1,200
4. Frankenstein, Gigantic 1:5 scale, Aurora, 1964	1,200
5. King Kong's Thronester, Aurora, 1966	1,000
6. Lost In Space, small kit, Aurora, 1966	900
7. Addams Family Haunted House, Aurora, 1964	800
8. Bride of Frankenstein, Aurora, 1965	750
9. Lost In Space, The Robot, Aurora, 1968	700
10. Godzilla, Aurora, 1964	500

But don't forget about vehicle model kits. Increased interest in the NASCAR circuit and drivers like Dale Earnhardt has placed a premium on related model kits. Revell-Monogram continues to be a leader in that field today. Other vehicle kits that look to have bright futures include Harley-Davidson motorcycles and possibly reissues of vintage 1960s funny cars by Tom Daniel and Ed Roth.

PEZ

PEZ collecting is growing at an incredible rate, said dealer Steve Glew.

"It's going ballistic," Glew said. "I would estimate that I get five to seven calls a week from people who are just starting in the hobby and want advice on where to go from here."

Glew estimated the popularity of PEZ collectibles has doubled in the last year.

In 1996, PEZ collectors hosted three conventions. The resulting media attention may be one of the reasons the hobby is so hot right now.

Among the most highly prized PEZ candy dispensers are Vucko, Mueslix and Make-A-Face variations. Also popular are Looney Tunes characters like Henry Hawk and Foghorn Leghorn.

Whenever Glew gets a call from a new collector his advice is always the same.

"I tell them to stop buying anything from anybody and start buying them over the counter at their local stores," he said. "Buy American product, then go to foreign product."

Once a dispenser has been off market for about six months, the price usually jumps about 30 percent, he said.

At this point, the Japanese have yet to enter the market. But if they do, Glew expects prices to double.

The Top 10 PEZ Dispensers

(in Near Mint condition)

1. Make-A-Face, American card	$3,000
2. Make-A-Face, German card	2,500
3. Mueslix	2,000
4. Witch Regular	2,000
5. Elephant	2,000
6. Pineapple	1,800
7. Make-A-Face	1,500
8. Space Trooper, gold	1,200
9. Lion's Club Lion	1,000
10. Advertising Regular	700

Restaurant Premiums

Collectors' appetites for restaurant premiums remained healthy, with Burger King toys among the preferred menu items. This is especially true for premiums from the mid-1980s and earlier.

McDonald's recently adopted a new generic display case to display future premiums. Since that time, demand for the old counter-top bubble displays has been high.

Demand for Happy Meal boxes also has picked up.

Top 10 Restaurant Premiums

(in Mint condition)

1. Big Boy Nodder, Big Boy, 1960s	$1,500
2. Big Boy Bank, Large, Big Boy, 1960s	300
3. Big Boy Bank, Medium Big Boy, 1960s	165
4. Colonel Sanders Nodder, Kentucky Fried Chicken, 1960s	150
5. Big Boy Board Game, Big Boy, 1960s	120
6. Transformers, McDonald's, 1985	110
7. Big Boy Bank, Small, Big Boy, 1960s	100
8. My Little Pony, McDonald's, 1985	80
9. Big Boy Stuffed Dolls, Big Boy, 1960s	80
10. Minnesota Twins Baseball Glove, McDonald's, 1984	75

The Top 20 Space/Science Fiction Toys
(in MIP condition)

1. Lost in Space Doll Set, Marusan/Japanese	$7,000
2. Space Patrol Monorail Set, Toys of Tomorrow, 1950s	4,000
3. Buck Rogers Roller Skates, Marx, 1935	3,500
4. Buck Rogers Character Figures, Britains, 1930s	2,500
5. Lost in Space Switch-and-Go Set, Mattel, 1966	2,300
6. Buck Rogers Uniform, Sackman Bros., 1934	2,100
7. Lost in Space Roto-Jet Gun Set, Mattel, 1966	2,000
8. Buck Rogers Cut-Out Adventure Book, 1933	2,000
9. Buck Rogers 25th Century Scientific Laboratory, Porter Chemical, 1934	1,600
10. Flash Gordon Home Foundry Casting Set, 1935	1,500
11. Lost in Space Chariot Model Kit, Marusan	1,500
12. Lost in Space Robot Model Kit, Aurora, 1966	1,500
13. Frankenstein Robot	1,400
14. Lost in Space 3-D Fun Set, Remco, 1966	1,200
15. Chief Robotman, KO, 1965	1,200
16. Buck Rogers Wristwatch, E. Ingraham, 1935	1,000
17. Godzilla Combat Joe Set, 1984	1,000
18. Buck Rogers Pocket Watch, E. Ingraham, 1935	1,000
19. Lost in Space Jupiter Model Kit, Marusan,1966	1,000
20. Lost in Space Jupiter-2 Model Kit, Marusan	1,000

Space/Science Fiction

As Kenner's Power of the Force figures and accessories made the jump to light speed, original *Star Wars* items from the 1970s and 1980s continued to hold their value. With the upcoming 1997 release of George Lucas's *Star Wars* trilogy — complete with added footage and enhanced special effects — and the scheduled 1999 release of a new *Star Wars* prequel, this line will likely stay hot.

The influence of baby boomers continued to assert itself in the area of space-related toys. The Japanese *Lost in Spac*e doll set topped the list of collectibles in this category. Other *Lost in Space* toys and several Buck Rogers collectibles rounded out the *Toys & Prices* top 20 list of space/science fiction toys.

Going where no science fiction collectors have gone before, many journeyed into the Barbie galaxy last year when Mattel released Star Trek Barbie and Ken as a duo. The set remains a hot item at *Star Trek* conventions nationwide.

According to dealer Pete Nyblom of St. Cloud, Minn., Playmates' *Star Trek* line of action figures zoomed along at warp speed, although the company's decision to release only 1,701 Captain Picard Tapestry figures angered many collectors who were unable to acquire one of the scarce toys.

Classic Tin

Prices stayed pretty much the same in this category with no real jumps from last year. Due to a busy market over the past few years, many of the top toys remained hard to find, which could drive up prices even further.

It's likely that more and more tin litho wind-ups and character toys will hit the market as collectors attempt to cash in on lucrative values.

Battery-operated toys will likely resurface, as well, continuing a trend that began in 1995. They are still fairly easy to find at toy shows, but collectors may now be on the lookout for those in better condition with clear and colorful original boxes.

The Top 25 Tin Toys
(in Mint condition)

1. Popeye the Heavy Hitter, Chein	$6,500
2. Popeye Acrobat, Marx	5,500
3. Popeye the Champ Big Fight Boxing Toy, Marx	4,600
4. Mikado Family, Lehmann	3,900
5. Red the Iceman, Marx	3,500
6. Snappy the Dragon, Marx	3,000
7. Popeye with Punching Bag, Chein	2,500
8. Masuyama, Lehmann	2,500
9. Lehmann's Autobus, Lehmann	2,500
10. Mortimer Snerd Band "Hometown Band", Marx	2,400
11. Hey Hey the Chicken Snatcher, Marx	2,400
12. Popeye Express, Marx	2,400
13. Hott and Trott, Unique Art	2,300
14. Superman Holding Airplane, Marx	2,250
15. Ajax Acrobat, Lehmann	2,200
16. Tut-Tut Car, Lehmann	2,200
17. Mortimer Snerd Bass Drummer, Marx	2,200
18. Chicken Snatcher, Marx	2,200
19. Popeye Handcar, Marx	2,000
20. Zig-Zag, Lehmann	2,000
21. Howdy Doody & Buffalo Bob, Unique Art	1,975
22. Ham and Sam, Strauss	1,950
23. Merrymakers Band (with marquee), Marx	1,900
24. Santee Clause, Strauss	1,895
25. Li-La Car, Lehmann	1,850

Vehicles

A close forerunner with action figures, vehicle toys remained the shining standard in toy collecting. While early prewar cast-iron continued to command the top prices, the most active collecting appeared to be in the postwar die-cast realm.

Topping this category in value were the classics from the 1920s to 1930s, including vehicles from Hubley, Arcade and Buddy L. Among the rarest of toy collectibles, a Mint condition Packard Straight 8 by Hubley can grab as much as $15,000.

Because of their scarcity and high prices, most collectors have turned their attention to smaller, more affordable die-cast vehicles, made by companies such as Ertl, Corgi, Tyco (Matchbox) and Mattel (Hot Wheels).

For instance, Matchbox cars are among the more affordable toy collecting investments one can make. The hot items of 1996 were the cars erroneously numbered #76, #77 and #78 in Matchbox's 1-75 series.

The Top 10 Corgi Vehicles
(in MIP condition; not including sets)

1. Man From U.N.C.L.E. Olds, cream version, 1966-69 .... $550
2. Popeye's Paddle Wagon, 1969-72 525
3. James Bond Moon Buggy, 1972-73 525
4. Batmobile, 1966 500
5. Green Hornet's Black Beauty, 1967-72 500
6. Chevrolet Performing Poodles Van, 1970-72 450
7. Beatles Yellow Submarine, 1969-70 450
8. Scammell Circus Crane Truck, 1969-72 450
9. Chitty Chitty Bang Bang, 1968-72 425
10. RAF Land Rover & Bloodhound, 1958-61 400

The Top 10 Mattel Hot Wheels
(in MIP condition)

1. Volkswagen Beach Bomb, surf boards in rear window, 1969 .... $5,500
2. Custom Camaro, 1968 2,000
3. Snake, 1973 1,500
4. Mongoose, 1973 1,400
5. Mustang Stocker, 1975 1,200
6. Carabo, 1974 1,200
7. Mercedes C-111, 1973 1,200
8. Custom Mustang, 1968 1,200
9. Ferrari 312P, 1973 1,100
10. Superfine Turbine, 1973 1,100

The Top 10 Vehicle Toys
(in Mint condition)

1. Packard Straight 8, Hubley $15,000
2. Checker Cab, Arcade 15,000
3. White Dump Truck, Arcade 15,000
4. White Moving Van, Arcade 13,500
5. Elgin Street Sweeper, Hubley 11,500
6. Ingersoll-Rand Compressor, Hubley 10,000
7. Yellow Cab, Arcade 10,000
8. Motorized Sidecar Motorcycle, Hubley 10,000
9. Borden's Milk Truck, Arcade 8,000
10. Ahrens-Fox Fire Engine, Hubley 8,000

Hot Wheels are "hot, hot, hot!" That's the word emphatically used by Don Potts of Don's Toys and Collectibles in Fort Worth, Texas. Limited-edition Hot Wheels, especially those in editions of 25,000 and less, sold particularly well. The Treasure Hunt limited-edition set of 12 cars was a particular favorite with collectors.

"It is almost impossible to get the new ones coming out, the demand is so great," Potts said. "The market is very strong. I think Hot Wheels are basically replacing the big push we had in sports cards.

The Top 10 View-Master Reel Sets
(Mint in Package; not including Movie Preview Reels)

1. Addams Family, #B486 .... $125
2. Munsters, #B481 125
3. NCAA Track & Field Championships, #B935 100
4. Dr. Who, #BD216 100
5. Dr. Who, #BD187 100
6. Davy Crockett, #935abc 75
7. Green Hornet, #B488 75
8. Dark Shadows, #B503 75
9. World Bobsled Championships, #B949 75
10. Lost in Space, #B482 75

Customizing Toys

By Guy-Michael Grande

If you want something done right, you have to do it yourself. So the saying goes.

Toy enthusiasts around the world are taking that saying literally when it comes to crafting their own custom creations. True, model kit bashers have done it since the dawn of their hobby. But today more than ever before, collectors are turning their creativity towards animating visions of their own with all kinds of toys.

More than vehicles or model kits, however, the largest area affected by this growing trend seems to be action figures, where fans of all skill levels are plying their talents into homemade heroes, heroines, fantastic creatures and villains. Some are familiar characters; others have never before been seen beyond their creators' imaginations.

Just how many people are out there customizing their own toys is unknown, but their ranks seem to be increasing. Some better known customizers include Scot Fleming of Pittsburgh, Pa.; Charlee Flatt of New Port Richey, Fla.; Bonnie Gilchrist of Edgewater, Fla.; Jim Marianetti of Tucker, Ga.; Polo Moreno of Davis, Calif.; and Dick Whittick of Clinton, Wash. Ask any one of them, and they'll attest to knowing fellow fans and friends who share an eye (and some talent) for customizing.

With so many toys available, both vintage and brand new, why the need to customize? Why do people do it? Several reasons are common among customizing circles. Some people customize to fit their level of quality. Some customize to create characters who have been omitted from established toy lines. Presumably, most

RETURN OF THE APES — These Planet of the Apes figures are actually customized G.I. Joe figures. Polo Moreno sculpts and designs his ApeJoe figures in his California home. Pictured are Cornelius and Zira (right), Dr. Zaius (below right), and Astronaut Taylor (below) four of the figures in his ApeJoe line.

DREAMING OF JEANNIE — Bonnie Gilchrist of Florida decided to customize her own I Dream of Jeannie *doll. Look familiar?*

customizers see what they do as providing an outlet for creativity and artistic expression which would be otherwise untapped. Those whose work has been seen by others — including the customizers mentioned here — have found ways to share their talents and inspire other collectors to follow suit, crafting characters to match their own visions.

The customizing process differs depending on who's behind it. It often involves considerable amounts of hunting, painting, prepping, sculpting, sewing and reworking, all the while keeping in scale for whatever they have in mind. Each customizer you'll meet here seems genuinely humble about his craft. Yet there can be no denying their talents, which are visible in nearly everything they've had a hand in customizing and creating.

Those who make the most minute of alterations on their dolls and figures are hesitant to call it customizing per se. Others take figures and redress them in handmade, handsewn fashions modeled after existing character costumes. Still others take customizing a step further, sculpting their own figures using only partially existing toy parts to form unique versions of favorite characters from comics, TV, movies and popular culture.

Why customize? To fill in the gaps, send in reinforcements and replacements and animate new visions of their own.

Filling in the Gaps

Scot Fleming says his drive to start customizing was simple.

"I've always been a big Batgirl fan, and there are very few Batgirl collectibles out there," he says. As far as action figures, there have been only three authorized renditions in the 30 years since Batgirl's debut — Ideal's 12-inch Super Queen from the Captain Action line (1967), Mego's 8-inch Super Gal from the Official World's Greatest Super Heroes line (1974) and Mego's 4-inch Batgirl Bendie (1970s). When Kenner's Super Powers line debuted in the mid-'80s, Fleming immediately noticed Barbara Gordon's alter-ego was missing from their well-detailed action figure ranks.

"She was one of the characters I thought they should've had," he recalls. "So I decided to try and make one for my collection." Using a Super Powers Wonder Woman, Fleming applied his modeling skills to create a Batgirl figure. Admired by enthusiastic convention-goers at an Ohio toy show soon thereafter, the figure would spark Fleming's custom works for years to come.

"So many people asked to see her and were so interested that it made me realize a lot of people share the same interests in figures that have never been done before," he observes.

Taking his cue from Batgirl, Fleming continued with customized versions of many female action figures who hadn't appeared in established manufacturers' toy lines. Not limiting his talents to that genre, he also followed his affinity for Adam West's TV Batman to create a Captain Action-style figure who bears an uncanny resemblance to West and a costume accurate down to the tiniest details.

"The hardest part actually ends up being the work itself," Fleming admits. "You know, prepping any given figure to turn it into another figure."

Like the majority of the customizers mentioned here, Fleming does not mass produce his creations. Because he does not pour or inject his own figures utilizing molds of any kind, Fleming uses existing figures, which he resculpts and repaints. As a result, he works in a scale much smaller than regular toy designers, who usually sculpt at twice or three times the size that ends up on the retail market.

Fleming says maintaining balance on that scale is essential.

"I try to keep it in scale with whatever line I'm working in," he explains. "I like the figure to look like the rest of the line." Size and articulation are important. If other figures in the line are jointed at the elbows and knees, his new figure has to be similarly articulated.

"That's the hard part, finding the base figures to do whatever you have in mind, and then trashing a few before you get the one that you like."

Trial and error is a theme that runs through every customizer's works. So far, Fleming's foibles haven't deterred him from applying his talents to an expansive array of toys and collectibles.

"I do all kinds of custom stuff, from illustrations to model-bashing to order," he elaborates. "It isn't all based on superheroes."

From his primitive first attempts to transform a G.I. Joe into Frankenstein, Charlee Flatt has been dedicated to making any superhero who hasn't yet materialized in action figure form and more. Inspired by his heroes, he allows that success hasn't always come easily.

Though Flatt, like Fleming, is steadily gaining widespread recognition for his craft, he's quick to note that not all of his creations came out so well when he started seriously customizing a decade ago.

Like so many other motivators behind toy collecting, Flatt believes his penchant for customizing was born in childhood. Rewarded for high grades with Mego figures during his school days, Flatt remembers the characters who got away. He's spent the past decade crafting figures that both kids and adults find enchanting.

"I'm a superhero fanatic, so I pretty much collect anything in a mask and spandex-type costume," Flatt admits, laughing. "But as far as customizing, as ridiculous as it sounds, if there is such a thing as fate or kismet, what I'm doing now must be mine, because I've been doing this my whole life."

An eternal fan of the Mego-style costumed superhero, Flatt's customizing process involves sculpting body parts and sewing costumes for his legion of 8-inch figures.

"When I first got started with my own costumes for the figures, I took apart a Batman costume, looked at how they sewed it together and duplicated it. That's the basic superhero costume — the top, the legs and the trunks.

"From there, I just customize. The sewing has always been really the easiest part to me; sculpting the heads is probably the toughest part."

Flatt has become adept at turning a tough process into a fine finished figure.

Batman's nemesis, Two Face, as created by Charlee Flatt.

"Originally, I would take a standard Mego head and grind all the detail off of it. Then I would put clay over on top of it and sculpt it into whatever face I wanted." Flatt says this process worked until his painstakingly-customized Mego heads expanded or melted beneath the clay faces he'd created, resulting in ruins.

"So I decided to figure out a way to make these heads entirely out of clay," he remembers. Flatt now begins sculpting heads with Super Sculpey clay.

"I'll sculpt the first head out of Super Sculpey, and at that point, make a slip-mold of it made of Mold Builder. It's a liquid latex you brush onto the surface of your item, let it dry and keep coating and coating until it gets thick enough to peel off. Then I pour a plaster copy." This copy is made using a gypsum cement called Hydrocal.

"Once I get a rough sculpting done, I'll make the mold and make a plaster duplicate. Then I'll work on the duplicate to make it what it want it to be," he says.

"I don't really have any secrets," allows the characteristically humble customizer. "I'm more than happy to share any techniques that I have with anybody."

Lucky for Mego collectors and superhero admirers, Flatt confesses he's "a total fan of 8-inch scale figures. I'm almost locked into this scale or something."

Besides filling out the ranks of known characters, making improvements on the original Mego-styled figures have come almost as a by-product of Flatt's process.

"I string 'em together differently than the Megos so that they'll hold poses better," he reveals. "I've just refined and improved as I went along. Making that decision to start sculpting the head completely myself just forced me into doing things a little bit differently, and from there, it kind of snowballed."

Garnering praise for his impressive creations, Flatt is quick to credit Fleming's influence.

"In so many ways, he was this gigantic inspiration to me. If there is an Elvis of toy customizing, it's definitely him. Scot's aces in my book."

Not every customizing gap is figure-based, however. Some customizers got their start with figures in need of new or different uniforms.

Focusing a trained eye toward the skies, Jim Marianetti knew something was missing in several service branches which had never been represented for G.I. Joe.

"I personally felt that Hasbro really didn't do enough for the Air Force," Marianetti explains. Recalling his father who'd flown B-26 bombers during World War II, he admits, "I would've been thrilled as a kid to have had a uniform similar to his." Realizing he was not alone in this opinion, Marianetti founded Aces, which has produced aviator uniforms for the past year.

For the company's authentically detailed B-17 Waist Gunner outfit, Marianetti tapped into Joe's historical roots. "We went to the original waist gunner on the Memphis Belle and asked him if he would sign certificates for us." The certificates, which contain the gunner's service photo and personal military background, were featured in a limited edition of 100 gift sets available for those customers who bought two or more Aces uniforms.

"We're really trying to key on Hasbro's promotion of the real American hero," Marianetti adds. "We're taking it one step further by representing real people, real history."

Aces soared to even greater heights when they produced their Flying Tigers and Panther Jet Pilot uniforms. "We went to the VF-111 Sundowners, because as we understand it, they were one of the first squadrons to get the Panther Jet. Their procurement officers sent us details on what the actual squadron wears — the insignias, the helmet design, everything. So we're trying to make everything as accurately as possible."

That quality has recently led to Hasbro recognition for Marianetti's Aces, with special limited edition boxed sets available through the G.I. Joe Collector's Club.

Reinforcements and Replacements

For Fleming, Flatt and Marianetti, their customizing inspiration is grounded in the toys they collect. Bonnie Gilchrist, on the other hand, will tell you she collects anything and everything.

"I love everything. I just have no favorites. I often think there must be something lacking in my personality," she says with a laugh.

Hardly lacking, Gilchrist got bit by the customizing bug when she started collecting antique dolls, realizing that those in Poor condition were far more affordable and available. Gathering a nursery full of them, she began her repairs in earnest and started crafting her aptly-dubbed "Celebrity Impersonator" dolls little more than a year ago.

Besides creating new figures on request such as a Teenage Pebbles Flintstone, Gilchrist also makes beloved characters like Samantha and Jeannie. Not duplicates of previous figural versions of the popular *Bewitched* and *I Dream of Jeannie* characters — both of which are incredibly pricey and rare on the secondary market — these two nonetheless provide collectors with stunning incarnations of Sam and Jeannie.

Working from color photographs, Gilchrist first finds a suitable doll and then crafts both expression and fashion to match.

"Well, I've always been artistic. I've always been crafty," she admits.

Her various creations range from the bizarre to the familiar, including Red Rider, The Girl from U.N.C.L.E., Lily Munster, The Green Hornet and his sidekick Kato, to cite a few. But while her dolls are statuesque and eye-catching, Gilchrist's basic stock does have one restriction.

"I do not use Barbies," she states unapologetically. "Barbie comes through everything with a big smiling face."

One of Charlee Flatt's creations, Mysterio

Photo and doll in hand, Gilchrist's typical routine involves painting the face, highlighting eyes, adding eyeshadow, eyelashes, perhaps changing the doll's mouth or adding teeth, and shadowing the jaws, all to create the desired expression. "Everything is done with either paints or chalks to get the look that I want. Then I seal it and put the outfit on it. I make a pattern and just make the outfit." One secret she won't share, however, is how she re-forms the doll arms to hold poses.

From start to finish, Gilchrist says her total working time often varies, depending on the character and how detailed their costume is.

"Maybe as little as two hours and many as eight or nine hours," she says. "The Green Hornet is a pest," she laughs, noting his multi-faceted outfit. "You're never done."

Gilchrist emphasizes that her creations are "Celebrity Impersonators," not knock-offs of toys already on the market.

"These are illusions. I'm not molding a doll; I do not sculpt the dolls. I might use one particular doll for six figures," says the illusionist.

Firmly grounded in reality, Cotswold Collectibles began making affordable G.I. Joe accessories when faced with the high-priced low supply of missing parts on every Joe collector's want list. Celebrating their first year and a half of business exclusive to their reproduction Joe line, Dick Whittick feels that affordability has been a key to Cotswold's ongoing success.

"We have sort of copycat rare pieces, but we also have a whole expanding line of things that were never made for G.I. Joe," Whittick explains. "It wasn't our plan to come out with such an extensive line," he admits. "The plan was to supply a few things to G.I. Joe collectors. But now it's just grown so that it is a whole separate business for us. There's a whole market for people who want accurate military weapons, uniforms, and so on."

Cotswold's expanding line has grown to include uniforms and other clothing accessories such as field jackets and cloth utility caps. Their finely- detailed replicas feature only the most subtle distinctions from original Hasbro accessories, causing Joe loyalists to press for even newer items to outfit their troops.

New Visions

While customizers have clearly been filling in the gaps and providing replacements for toy collectors, some of them have given new life to the hobby with innovative new characters and designs.

Like the legendary Reeses Peanut Butter Cup, Polo Moreno has combined two popular fictional character lines to the sheer delight of fans everywhere with ApeJoe, the perfect blend of *Planet of the Apes* and G.I. Joe.

"I took my two favorite things and put them together," Moreno says, chuckling. The California- based state biologist claims he started making the 12-inch ApeJoes "to relax." But with the 30th anniversary of *Planet of the Apes* fast approaching, collectors are bound to keep him far from relaxed.

Utilizing 12-inch reproduction Joe bodies made by Cotswold, Moreno sculpts intricate molds for each ApeJoe head and fashions their clothing by hand. His eye for detail is astonishingly accurate, from General Urko's menacing glare to Dr. Zaius' skeptical brow. Designing the head for Kim Hunter's leading character Zira took three separate attempts before Moreno had her expression where he wanted it. Like a seasoned Hollywood makeup artist, he knows the proof lies in every crease and wrinkle.

Moreno's growing ApeJoe ranks include Taylor, Zira, Zaius, Urko and gorilla soldiers. A year old, these fully poseable figures are expected to be joined by the mysterious masked mutants and other *Planet of the Apes* characters in the near future.

Everywhere you look, custom figures and toys are turning up at toy shows, catching the eyes and rapt attention of collectors. The possibilities are as endless as each customizer's imagination.

"Some people smoke. Some people drink coffee. Some people customize action figures," offers Flatt.

No one could question that customizing gives us missing characters and a great outlet for creative expression. But still a nagging thought persists — why would anyone want to customize, say, a Toy Biz figure into Batman when there are gazillions of Kenner Batman figures riddling toy store shelves and secondary market bins alike?

"Because the Kenner ones are pretty much Starting Lineup," Flatt replies. "Arm up, arm down. To me, that's an 'inaction' figure."

One of the visible results of the growing customizing hobby is currently taking place at the level of toy manufacturers themselves. Some companies are becoming more responsive to these issues by including more diverse characters, female action figures, and improving their own details and quality.

For some, like Cotswold Collectibles, customizing has become a business. For others, it's something they enjoy doing in their spare time. For all toy collectors, customizing is an art form to enjoy and look forward to, maybe even an inspiration to make some of your own.

"There are a lot of people out there who customize their own," says Fleming. "But I also meet people who are just not good with anything. The''re not artists, they're not craftsmen. They're not model builders at all. They don't have time for this. And to them, it's more of an amazing miracle than it is to fellow hobbyists."

"I think that a lot of people just see characters that they feel they need to have," Flatt observes.

"They look at them and immediately want them to fill up their lines, because it's a character they've always loved that no one's ever made," Fleming agrees. "There's always someone out there that whatever obscure character you've made, that's their favorite character."

"I couldn't really tell you what it is that brings people to doing this," admits Flatt, a touch of humor in his voice. "I wish I knew. I fear that it's some kind of mental disorder," he laughs.

Contributing writer Guy-Michael Grande is unsure how he regularly customizes his own words into feature articles and his Shop Talk column for Toy Shop.

Think you might like to try your hand at customizing action figures? Here's how to reach the customizers mentioned in this story.

Scot Fleming
Custom Works
843 Rocklyn Dr.
Pittsburgh, PA 15205
412-734-8424

Charlee Flatt
B Natural Toys
5800 Tennessee Ave.
New Port Richey, FL 34652
813-841-9121

Bonnie Gilchrist
B. Gilchrist Antiques
2926 Mango Tree Dr.
Edgewater, FL 32141
904-423-8921

Jim Marianetti
Aces
P.O. Box 496
Tucker, GA 30085
770-934-5652

Polo Moreno
ApeJoe
920 Cranbrook Ct. #17
Davis, CA 95616
916-758-9532

Dick Whittick
Cotswold Collectibles
P.O. Box 249
Clinton, WA 98236
206-579-1223

When Old is Really New

By Mike Jacquart

Reproductions. The word probably sends shudders down the backs of many toy dealers, but reproduction toys can have benefits.

As Mint and Mint in Box originals become scarcer and more expensive, reproductions can ensure that toy collecting remains a hobby for the masses, not just the wealthy.

Some companies, recognizing that not everyone's pocketbook matches his eagerness for the hobby, are manufacturing dolls, cast iron, still and mechanical banks and other reproductions of original toys.

Rocky Sorrentino is one person in the toy industry who helps fill this niche. His company, Art 'n Things, specializes in recreations of classic Aurora model kit boxes. Sorrentino takes painstaking detail to duplicate the packages and artwork of the company's monster, superhero and other figure kits. Sorrentino, whose repro boxes range from Frankenstein to John F. Kennedy, believes his company serves a vital function in the toy trade.

"Reproductions are more within the realm of affordability for hardworking, middle class people," Sorrentino explained. "[Also] some people have the built-up kit, but they've lost the box over the years."

Aurora Repros are Favorites

Other companies share Sorrentino's view that reproductions are a boon for collectors who can't afford original toys and for people who want to play with their toys but don't want to risk damaging valuable collectibles.

Playing Mantis, under the Polar Lights trademark, has created nearly identical versions of two popular Aurora kits, the Addams Family Haunted House and the Mummy's Chariot. Like the originals, the Addams Family house is licensed from Paramount and is based on the television show of the same name. The Mummy's Chariot, like the 1965 kit, is a blend of a haunting character and a monster vehicle.

Playing Mantis President Thomas Lowe built both models when he was a boy, and the Addams Family Haunted House was one of his favorites.

"I wanted to remake these kits because they were originally made with such creativity and craftsmanship," he said. "I wanted every model kit maker to have the opportunity to build one without destroying a kit valued in the hundreds."

An original, Mint in Box Addams Family Haunted House can bring $800, and a MIP Mummy's Chariot can sell for $450. The repros are priced at $20 each.

The Louis Marx Company has also gotten in on the act. The company, an industry giant in the 1950s and 1960s, is reissuing some of its best-selling play sets, including Fort Apache Stockade, the Alamo, Battleground and Cape Canaveral.

"We own the original molds," said Marx Toys President Jay Horowitz, "and we're using the original molds, stands and materials for these sets."

Fort Apache Stockade, like the play set issued in 1953, includes a set with plastic fences ($24.95) and a more expensive version ($49.95) with tin lithographed fences.

"Most [original] sets had plastic fences, but some had tin," Horowitz said, adding that 34 variations of the set were made between the 1950s-1970s.

The repros, like the originals, feature over 100 pieces including cowboys, Indians and of course, the fort.

The Alamo, with its tin lithographed buildings, cannons, ladders and U.S. Cavalry and Mexican soldiers, also includes the same pieces as the original. At $99, the Alamo is the most expensive set, but Horowitz pointed out it's still reasonable compared to the originals. A Mint in Box Alamo set can bring $500 or more, and a MIB Fort Apache Stockade set from '53 is valued at $255 or more.

In addition, Horowitz said the sets aren't any more expensive for today's collectors than they were for fami-

SEEING DOUBLE? — The Marx Toy Company is currently reissuing some of its popular 1960s play sets. At left is an original Davy Crockett at the Alamo set. Note the Walt Disney name at top left. The company logo at bottom right is also slightly different than the original. In this case, the repro piece is obvious, but sometimes reproduction toys and packages are more difficult to spot.

lies in the 1950s and 1960s.

"They cost $5.98 originally," he said, "and, accounting for inflation, that's about the same as $50 today."

While Marx has strived to make the repros nearly identical to originals, it's still obvious the sets aren't the real thing, Horowitz said.

"Each [set] will include the date it was issued and a booklet containing histories of both the Marx company and the period in which the toy was set," he said.

Small Business vs. Big Business

Reproductions still have raised concerns. Do they keep dealers from charging higher prices for originals? Will dealers be run out of business if enough repros are produced?

Florida dealer Bonnie Gilchrist produces and markets her own versions of dolls like Ideal's Batgirl, but she evidently sees repros as a double-edged sword — they help fill a niche, yet they may also drive down prices of originals in the long run.

"Why would someone pay $500 for a Marx gas station if they can buy one just like it for $30?" she said.

As more large toy companies start making reproductions, independent toy dealers will feel even more threatened, according to Gilchrist.

Other Ethical Concerns

The threat of big businesses squeezing small dealers isn't the only concern some people have about reproductions. While dealers believe most of their peers are honest, some people undoubtedly can, and have, tried to pass off reproductions as originals.

Gilchrist said she's seen "cast-iron" Coca-Cola trucks come into her shop that are actually forgeries from China. Some dealers will tarnish cast-iron toys to make them look older, take them to shows and try to pass them off as originals, she said.

Unscrupulous dealers like these hurt the industry, she said. "I know people who wouldn't touch cast iron because of things like that."

Bob Peirce, who holds still and mechanical bank auctions in the Midwest, said he's known people who have buried banks in their yard to make them look old.

"They get rusty and the paint flakes off," he said. "Paint in Excellent condition is what makes a 100-year-old bank valuable. If it's missing any more than a little paint, you shouldn't be buying it whether it's original or not."

It's also possible for dishonest dealers to use repro parts to finish an otherwise incomplete original. Repro kits used to be fairly expensive to produce, so it didn't make sense for a shady dealer to spend $200 revamping a repro that was only worth $100 built up to begin with. But as costs dropped, temptations rose to use gimmicks to fool collectors — in essence, potential profits from forgeries have, unfortunately, never been better.

While it's probably not possible to guarantee that your $400 Aurora model kit isn't really an inexpensive repro, Gilchrist believes most dealers are trustworthy.

"I think most people want to believe they're getting the original parts, not brand new items," she said.

Know the Difference

Luckily for collectors, it's possible to tell at least some repros from originals. Today's imitation model kits use a harder resin than their predecessors, which means a new kit sould weigh more than the real thing.

Cast-iron imitations can be detected as well, Gilchrist said. Vintage cast iron was very smooth with numerous, heavy coats of enamel.

"New cast-iron pieces aren't as smooth, and they're often not done with as many coats of enamel," she said.

Many, but not all still and mechanical bank reproductions, like those of the popular Trick Pony and Trick Dog, are marked as reproductions, Peirce said.

"If there's an indented circle on the bottom about the size of a half dollar, it's a reproduction," he said.

What if there is no mark? Feel and size are other important distinctions in still and mechanical banks, Peirce said.

"Old banks are smooth while repros will usually have a rough feel," he said. "Repros are also somewhat smaller than an original. It may only be an eighth of an inch, but it's still significant.

"Overall quality is also better than repros," Peirce added. "Sometimes it's subtle things like more defined lines and sharper paint, but they're things that can be detected."

But that doesn't mean all repros are shabbily made — far from it, according to Horowitz and Sorrentino. Horowitz said a great deal of craftsmanship has gone into his company's play set reissues to ensure that the baby boomers who bought them decades ago can pretend they're playing with the same set.

Sorrentino uses a color laser printer to stockpile a large selection of lithographs produced from the original boxes. He assembles every box by hand. He uses heavy-gauge cardboard and takes great care to ensure each box is the same size and shape as the originals. He also glues the lithograph to the cover and sides and seals the box in protective shrink wrap.

"It's a rigorous process," he said, "but you've got to make them look like the real thing even if people know they're not. That's what people are paying for, the illusion that they are originals."

Collectors shouldn't rely on dates to determine authenticity either, dealers said. For instance, Shirley Temple dolls made in 1972 are still being made today in China with dates from that year on the back. Until recently, Barbie dolls contained a 1960s copyright. Why is this done? The dates are copyright dates, not manufacturing dates, but they can be confusing, especially for novice collectors.

"It's important for beginners to not rely on dates to identify items," Peirce said.

Still, undoubtedly some dealers will try to take advantage of the confusing dates to pass off newer items as being much older than they really are, he said.

Be Honest With Customers

Gilchrist said it's important for dealers to make sure collectors know they're buying a reproduction. She said she makes it clear to her customers when an item is a lookalike and when it's an original. That's important to eliminate any confusion between her products and the real thing, she said.

Sorrentino also takes great steps that leave no doubt his work is an imitation. His name, company name and phone number are listed on all four inside flaps. Sorrentino believes it's only fair reproductions be labeled as such.

"We know some people are going to try to pass it off as the real thing," he said.

Gilchrist agrees.

"My dolls are always marked 'lookalikes' on the box," she said.

Other Advice

Collectors should also be aware that reproductions exist in tin, action figures and other types of toys — not just model kits, cast iron, banks and play sets.

Maybe the best tip regarding reproductions is a nugget of advice you might have heard before — only do business with dealers with solid reputations. Ask questions and do your research to make sure you're a knowledgeable collector.

They Say It's Your Birthday!

Milestones in Toy History

Did you know Lincoln Logs, a popular building toy, were developed by the son of the famous architect Frank Lloyd Wright? Were you aware that 700 million Play-Doh brand modeling compounds have been sold since its inception?

These were among the major toy anniversaries observed in 1996. Take a trip down memory lane, and you'll probably recall some of the fun you had playing with the toys listed below.

120 Years

Tootsietoy

Tracing its roots back to 1876, Strombecker, manufacturer of Tootsietoy die-cast vehicles, is one of the oldest toy companies in the United States. Tootsietoy vehicles have included cars, trains, trucks, airplanes and spaceships. Other Tootsietoy products include trains, bubble toys, action figures and jump ropes.

The company celebrated its 120th birthday in 1996 with new die-cast toys including Hard Body vehicles. The company was named after Toots, one of the founder's granddaughters.

This Rambler Station Wagon pulled a trailer. Both pieces are vintage Tootsietoys.

100 Years

Louis Marx's Birthday

Louis Marx, one of the most creative toy makers of all time, was born in Brooklyn, N.Y. in 1896. Marx played a major role in making well-made toys available to the masses, not just the wealthy. The Amos 'N Andy 12-inch walker, Fresh Air Taxi, Imaginative Crazy Cars and Jumpin' Jeep are but a few of the tin lithographed vehicles produced by Marx's company.

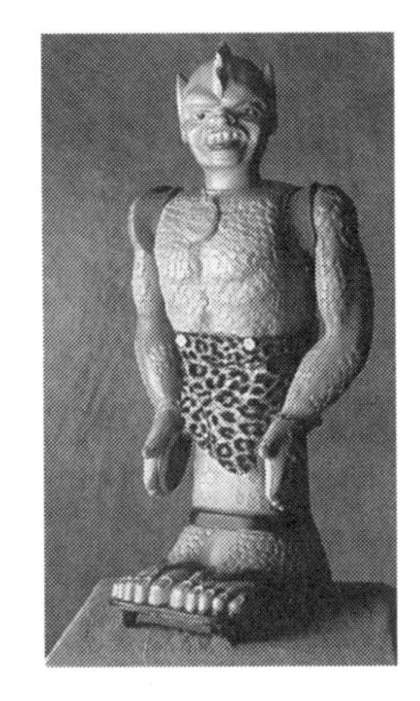

The Great Garloo, a popular battery-op toy by Marx.

Other notable toys have included play sets (Fort Apache, Alamo, Cape Canaveral), guns (Tommy Gun, Army Automatic Pistol) and robots (The Great Garloo, Rock 'Em Sock 'Em Robots and Big Loo). Marx didn't live to celebrate his 100th birthday, but the company, which has been sold numerous times, still produces toys that are putting smiles on the faces of children and adults alike.

80th Anniversary Edition Frontier Town Lincoln Logs

80 Years

Lincoln Logs

No one knows how many engineers and architects this toy may have spawned, but Lincoln Logs, a notable building toy, celebrated its 80th anniversary in 1996. Playskool noted the occasion by issuing a limited-edition frontier town. The 237-piece set, which included the original redwood-colored logs, allowed youngsters to create Wild West buildings like a fort, barn, cabin and general store.

50 Years

Tonka

These trucks have been tough enough to last 50 years. Tonka, known for its durable, large scale die-cast cars and trucks, was originally known as the Mound Metalcraft Company in 1946. Tonka has seen many changes over the years, including Tiny Tonkas, space

shuttles, helicopters and its purchase by the Hasbro Toy Group in 1987. But its slogan, "A toy shouldn't break just because a kid plays with it," continues to this day.

Nylint

Nylint officials like to keep a low profile, preferring to concentrate on doing a good job on its usual die-cast vehicle lines instead of making a fuss over its 50th anniversary in 1996. Nylint's high-quality toys have included construction vehicles, vans, street sweepers and hook 'n ladder fire engines. In today's plastic age, the company prides itself on continuing to also use steel to make its toys. Nylint's name is derived from company founders David Nyberg and Barney Klint.

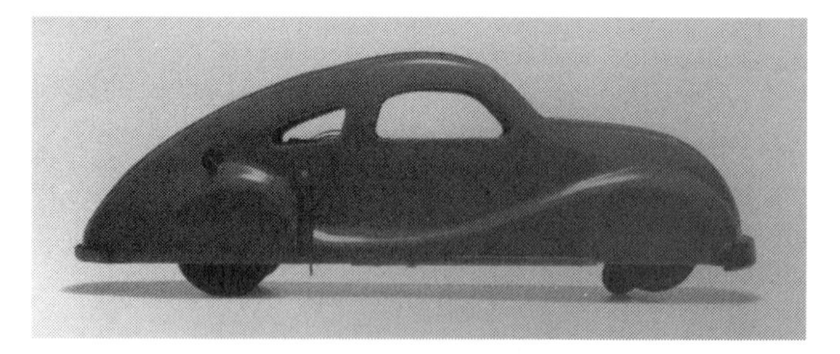

Nylint's first toy car was manufactured in 1946.

45 Years

Colorforms

Despite its modest beginning on a bathroom wall, Colorforms have become classic toys and valued collectibles. The colorful boxed play sets with unique stick-on plastic pieces were the brainchild of New York artists Harry and Patricia Kislevitz. The couple was looking for a way to design a collage that could be pieced together and removed when they created the first Colorforms — a brightly-colored canvas in their bathroom. The first Colorforms play set appeared in stores in 1951. The company went on to produce many play sets, including a limited-edition (20,000) version of its original set in 1996.

Colorforms were first introduced in 1951. Pictured is an early Captain Kangaroo's Treasure House Game.

40 Years

Gumby

Created by Art Clokely, Gumby made his first appearance in 1956 on *The Howdy Doody Show*. Since then, the little green clayboy, his pal Pokey and other friends have appeared in cartoons, a movie and have taken shape as numerous toys as well as T-shirts, mugs and other items. Gumby is fun, flexible and even-tempered — he never gets bent out of shape.

Gumby and Pokey were created 40 years ago.

It's uncertain whether Play-Doh will become a collectible, but it has been delighting children for 40 years. Pictured is one of Playskool's more recent play sets.

Play-Doh

If you've ever enjoyed playing with this moldable toy, you have Joseph McVicker to thank. McVicker, of the Cincinnati-based Rainbow Crafts Company, introduced Play-Doh in 1956. Originally only available in white, Play-Doh brand modeling compounds have since been offered in almost every color under the sun, including scented clay in colors like Splurple and Pinktastic. Its secret formula remains a mystery today.

Yahtzee 40th Anniversary Collector's Edition by Milton Bradley

Yahtzee

A Canadian couple designed the Yacht Game in 1956 to give their well-to-do friends something to do aboard their yacht. Much to their chagrin, they signed away their rights to Edwin Lowe in exchange for a few copies of the game. Lowe, who made a fortune selling bingo games in the 1920s, changed the name of the now classic dice game to Yahtzee. Milton Bradley bought the Lowe Company and the game in 1973. Yahtzee's slogan is, "The Game That Makes Thinking Fun."

35 Years

The Great Garloo

The Great Garloo. Friend or foe? Monster or buddy? Whatever you think of the Great Garloo, Marx's 2-foot windup toy introduced in 1961, there's no doubt the great gargantuan is a "monstrous" collectible for some fans of science fiction toys.

Barbie Queen of the Prom Game

Buy a dress, get a boyfriend (or is that the other way around?) and become queen of the prom and you're the winner of this game — Mattel's Barbie Queen of the Prom Game, which celebrated its 35th anniversary in 1996. To commemorate the event, Mattel issued a reproduction of the ultimate Barbie board game.

Frosty Sno-Man Sno-Cone Machine

What baby boomer didn't like to make his own sno-cones? The Frosty Sno-Man Sno-Cone Machine, introduced by Hasbro in 1961, included all the crushed ice, cups and syrups needed for youngsters to make 10 different sno-cones. A deluxe set later included a vendor's hat and apron.

Talking Beany and Cecil

"Help Cecil, help," pleaded Beany Boy. "I'm coming, Beany," countered Cecil, Beany Boy's sea serpent friend. Those are among the phrases recited by Mattel's Talking Beany Boy and Talking Cecil. The dolls were among the Beany and Cecil toys Mattel produced in 1961 in conjunction with the popular animated children's series that featured the lovable, blond-haired lad and his large green friend.

30 Years

Captain Action

Captain Action, unveiled by Ideal in 1966, was a true master of disguise. The action figure, designed to counter the popularity of G.I. Joe, had numerous alter egos including Flash Gordon, Spider-Man, Aquaman, Batman, the Lone Ranger and Superman. Dr. Evil was created as Captain Action's archenemy. While Captain Action didn't

LEFT: Barbie Queen of the Prom Game.
ABOVE: Captain Action by Ideal, 1966.

last long on the toy market, it's a valuable collectible today.

Francie

Francie, Barbie's mod cousin, is still looking hip after 30 years. A reproduction of the original 1966 Francie doll was issued in 1996. Francie's attire included her two-piece swimsuit and GadAbout, a popular outfit that featured a knit sweater and skirt, patterned stockings, knit cap, shoes and go-go glasses.

Mattel's Francie was reissued in 1996 for her 30th anniversary.

Batmania has lasted over 30 years!

Everybody's favorite party game — Twister — turned 30 in 1996. Has it really been that long?

Twister

This game had nothing to do with bad weather. Twister, unveiled by Milton Bradley in 1966, was the game that tied you up in knots. The game was popularized by *Tonight Show* host Johnny Carson and guest Eva Gabor in '66. Seeing how much fun Carson and Gabor had playing the game, Americans rushed to stores to buy their own copy of the game.

Lost in Space Robot

He may not have had a name, being referred to simply as "Robot" or "The Robot," but with phrases like "Danger! Danger!" and "That does not compute," the *Lost In Space* robot is still one of the most famous tin men of all time. Remco's toy version, produced in 1966, traveled along the floor and could be programmed for turns.

Star Trek

Star Trek isn't a toy, it just seems that way. Trek toys were in abundance in 1996 to mark the 30th anniversary of the original *Star Trek* series. Toys included Playmates' action figures of original Trekkers Captain James Kirk, Yeoman Janice Rand, Nurse Christine Chapel, Lieutenant Uhura and a replica of Dr. McCoy's medical kit. Considered the biggest non-Disney enterprise, *Star Trek* has lived long and prospered.

A scene from Galoob's 1996 Star Trek line.

Spirograph

Compasses weren't fun for children until Kenner introduced Spirograph design toys in 1966. With compass and marker in hand, the number of designs that could be made were as limitless as a child's imagination.

Remember how popular the Green Hornet was in 1966?

Batman, Green Hornet

Numerous *Batman* and *Green Hornet* toys were produced in 1966 in conjunction with the television series debuts of both superheroes. Batman toys from '66 included an Aladdin lunch box and Thermos bottle, Ideal hand puppet, Aurora Batmobile model kit and Corgi Batmobile. Green Hornet toys introduced 30 years included a Whitman coloring book, Ideal hand puppet, PEZ candy dispenser, Aurora Black Beauty slot car, Milton Bradley Quick Switch Board Game and Colorforms' Print Putty and cartoon kits.

25 Years

UNO

UNO celebrated its 25th anniversary in 1996. Mattel has issued numerous variations of the game since its inception — including UNO Madness, UNO Hearts and UNO Dominoes.

"You'd best listen up, Private. If you collect G.I. Joe, Action Man, The Elite Brigade or any other uniformed figures you **will** ask Cotswold Collectibles for their catalog. Do I make myself clear, Private?"

Cotswold Collectibles

Home of One-Stop G.I. Joe Shopping

PO Box 249, Clinton, WA 98236 360/579-1223 Fax: 360/579-1287

Free Illustrated Color Catalog: GI Joes from 1964-1996

Visit our Web page at http://www.whidbey.net/~cotswold

WANT IT BAD? NOBODY MAKES IT? WE DO!

A.F.E. has been creating and customizing figures for over five years!

Most work done within one week!

Very low prices!

We have ads every month. . . so see what's new!

STOP wishing you had it. . . Call us and get it!

Everybody wants something;

SO CALL OR WRITE NOW!

ACTION FIGURE EMPORIUM

604 Fairwood Drive
Nashville, TN 37209
Attn. Robert Walker
1-615-356-3170

Toy locating and model assembly services available also! Call or write today!

Buying plastic model kits from the 1950s-70s, Airplanes, Missiles, Figures, Sci-Fi, Gift sets, Military, Automotive, Ships. One kit or entire collection. All manufacturers.

Also buying Non-Sport cards from the 1950s-70s, set and singles.

Will pay premium prices for model gift sets, original factory displays & original artwork.

CREATIVE THREE HOBBIES

JIM BANKO
7475 Miller Drive
Bath, PA 18014
Phone: (610) 837-8582
FAX: (610) 837-4507
Email: EWPG79A@prodigy.com

MATCHBOX — Hot Wheels

Specializing In Original Die-Cast

Dan Wells ANTIQUE TOYS

CORGI — JOHNNY LIGHTNING

All Manufactures

☐ *YES! Send me the next four quarterly issues of Dan Wells Antique Toys Toy Catalogue, listing over 5,000 toy cars for sale.*

Name ____________________

Address ____________________

City ____________ State ____ Zip ____

Signature ____________________

Credit Card No. ____________________

Expiration Date _____/_____ Mail To: **Dan Wells Antique Toys**

7311 Highway 329 Ste. 601
Crestwood, KY 40014

☐ Sample $4.00
☐ Bulk Mail $12.00
☐ Priority $24.00

☐ UPS Next Day Air $60.00 (street addresses only)
☐ Outside USA Air Mail $24.00 (including Canada)

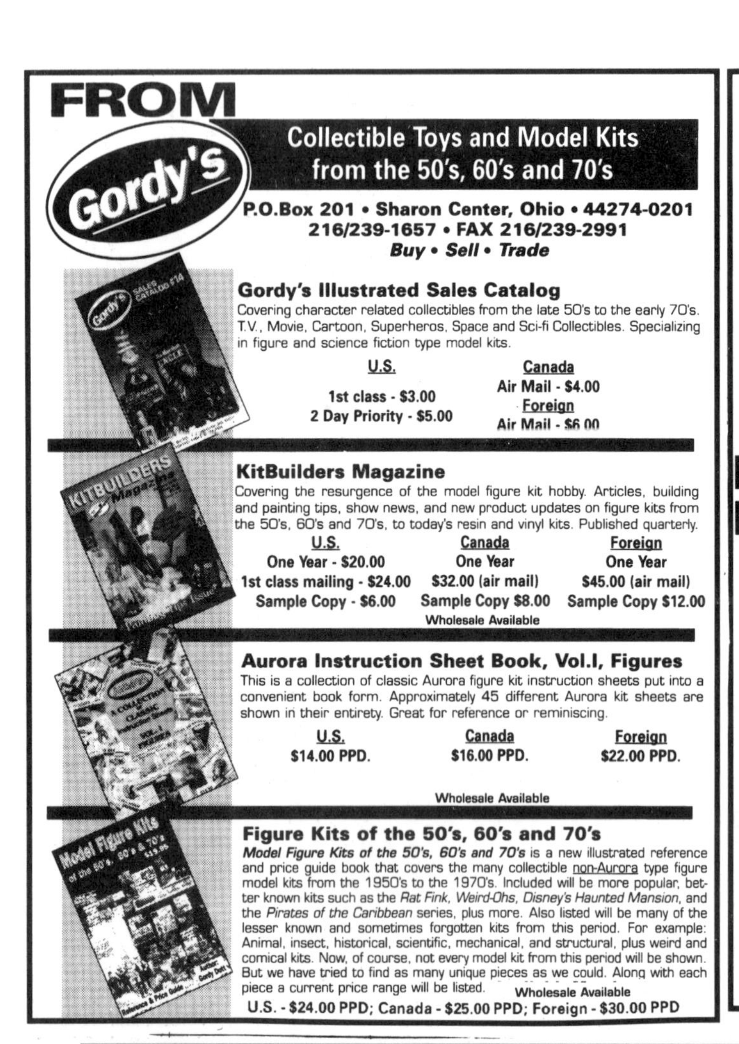

THE EARTH IS BUYING KENNER

TOYS

We sell collector supplies... stands for loose figures and clear zip-lock bags in all sizes

(513) 561-TOYS

SPLASH PAGE COMICS & TOYS

Serving Collectors Since 1986

We carry:

American Toys
Japanese Animation
Collectible Card Games
Role-Playing Games

Original Animation Cels
Japanese Toys
Videos

Trading Cards
Models
Applause

And Much, Much More!

International Orders Welcome!

Splash Page Comics & Toys
P.O. Box 158, Kirksville, MO 63501-0158

Retail Store: 800-237-7243
Secondary Number: 816-665-0188
Intl. Phone: 816-665-7623
FAX: 816-665-1657
E-Mail: SplashPage@Aol.com

And for the most current info on our in stock items check out our web page at http://www.splashpagecomics.com

PAST-TIME HOBBIES, INC.

9311 Ogdin Ave.
Brookfield, IL 60513

(708) 485-4544 / Fax (708) 969-8959

SPECIALIZING IN THE VERY BEST IN COLLECTABLE CARS, TRUCKS, FIRE VEHICLES, MILITARY VEHICLES & MORE!

SCALES: 1/87 THRU 1/12th PLUS VARIOUS EXCLUSIVE LIMITED EDITIONS

RETAIL STORE & MAIL ORDER

WE SHIP WORLDWIDE
WAREHOUSE FINDS • ONE-OF-A-KINDS
"Just 200 feet from the World Trade Center"
NOSTALGIA • SPORTS • MOVIE • T.V. • ROCK & ROLL • POLITICAL • SCI-FI • COCA-COLA • ADVERTISING • MEMORABILIA • VINTAGE TOYS • VINTAGE JEWELRY • CARDS • COMICS • JUKE BOXES • VENDING MACHINES

18 VESEY ST.
New York City, NY 10007
(212) 267-7100

"The Best In The Business"

STAR WARS
Pepsi-Cola

STORE HOURS:
Mon.-Thurs. 9 a.m.-7 p.m.
Fri. 9 a.m.-7 p.m.; Sat. Closed
Sun. 11 a.m.-6 p.m.

GIANT CATALOG AVAILABLE $2.00
E-mail WCC@earthlink.netearthlink
Home Page http://home.earthlink.net/~wcc

Bob's Toy Car Show
"Limited Edition Collector Cars"

Johnny Lightning 1954 Chevy Panel
Honolulu blue with yellow flames
and
Bob's Toy Show logo in a red & yellow banner

$15.00 each

plus shipping
Buy 12 or more - $12.00 each • Quantity discounts available
Information or to **order** call, FAX or write
Bob Goforth - Bob's Toy Show
4061 East Castro Valley Blvd. Suite 224, Castro Valley, CA 94552
(510) 889-6676 • FAX (510) 581-0397
All cars roll on Cragar SST wheels.

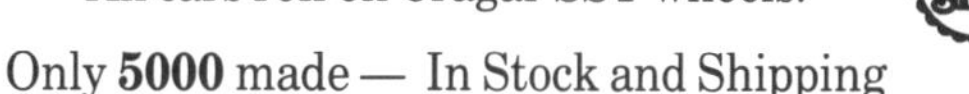

Only **5000** made — In Stock and Shipping

Bob's Toy Car Show
"Limited Edition Collector Cars"

Hot Wheels Delivery Van #13

Red and Yellow Van
and
Bob's Toy Show logo in a yellow banner
with red lettering

VAN ROLLS ON REAL RIDER TIRES

$15.00 each

plus shipping
Buy 12 or more - $12.00 each
Information or to order call, FAX or write
Bob Goforth - Bob's Toy Show
4061 East Castro Valley Blvd. Suite 224, Castro Valley, CA 94552
(510) 889-6676 • FAX (510) 581-0397
Only 8,000 made - in stock and shipping!

Directory of Notable Toy Manufacturers

Past and Present

Have you ever wanted to reach a toy manufacturer but didn't know how? Or maybe you wanted to know if a particular company was still in business?

While not all-inclusive, this list does include many of the notable toy companies from the 1800s to the present. Some companies are long extinct, but have been listed to give collectors a historical perspective on the industry. Toy manufacturers, especially historic ones, often had sparse or non-existent records and archival information. Companies without a current address and phone number are either defunct or listed under another cross-referenced company name.

We have tried to ensure the accuracy of every entry, but if you know of a change or correction that should be added to our database, please write to Sharon Korbeck, editor; *Toy Shop*; 700 E. State St., Iola, WI 54990.

Company Name	Product	Further Information	Current Address	Phone
Akro Agate	Marbles/Glass	Clarksburg, WV, 1930s-'40s		
Aladdin Industries	Lunch Kits		703 Murfreesboro Rd., Nashville, TN 37210	615-748-3000
Alexander Doll	Dolls	1923-present; founded by Madame Beatrice Alexander Behrman	615 W. 131st St., New York, NY 10027-7982	212-283-5900
All Metal Product Co. / Wyandotte Toys	Miscellaneous	Wyandotte, MI, early to mid 1900s, produced metal games and toys		
All-Fair	Games	New York locations, 1920s-1950s		
Applause	Plush, PVC figures	merged with Dakin in 1995	6101 Variel Ave., Woodland Hills, CA 91365-4183	818-992-6000
Arcade Mfg. Co.	Cast-Iron Vehicle Toys	Freeport, IL, 1885-1946, sold at end of World War II		
Auburn Rubber	Miscellaneous Rubber Toys	Auburn, IN 1930s-1960s		
Aurora	Miscellaneous	1961-1970s, produced figural and vehicle model kits, slot cars and sets		
Bachmann Industries	Toy Trains, Planes		1400 E. Erie Ave., Philadelphia, PA 19124	215-533-1600
Bandai America	Action Figures		12851 E. 166th St., Cerritos, CA 90701	310-926-0947
Barclay	Cast-Iron Vehicle Toys	Moline, IL 1920s-1960s		
Binney & Smith	Crayola Products, Silly Putty		1100 Church La., P.O. Box 431, Easton, PA 19042	610-253-6271
Bliss Mfg. Co.	Games	Pawtucket, RI, 1870s-1915, produced lavishly-illustrated Victorian parlor games		
Brio	Miscellaneous		6555 W. Mill Rd., Milwaukee, WI 53218	414-353-8677
Britains, LTD.	Toy Soldiers	1893-1950s, British producer of metal soldiers; now distributed by Reeves International		
Buddy L	Vehicle Toys, primarily trucks	filed Chapter 11 and sold to Empire Industries in 1995		
Cadaco	Games	founded 1935, previously known as Cadaco-Ellis	4300 W. 47th St., Chicago, IL 60632-4477	312-927-1500

Company Name	Product	Further Information	Current Address	Phone
Chein Industries	Tin and Mechanical Toys	New Jersey, 1903-1978		
Cohn, T., Inc.	Miscellaneous	New York, NY, 1950s-1960s		
Collegeville / Imagineering	Costumes		4th & Walnut Sts., Collegeville, PA 19426	215-489-0100
Colorforms	Colorforms Play Sets		133 Williams Dr., Ramsey, NJ 07446	201-327-2600
Corgi Classics	Die-Cast Vehicle Toys	distributed through Reeves International in U.S.	Meridian West, Leicester, LE3 2WT, United Kingdom	
Craft House Corporation	Miscellaneous		328 N. Westwood Ave., Toledo, OH 43607-3343	419-536-8351
Dakin	Plush, PVC figures	1955-present; merged with Applause in 1995	20969 Ventura Blvd. #25, Woodland Hills, CA 91364	818-340-5800
Duncan Toys	Yo-Yos		15981 Valplast Rd., Middlefield, OH 44062	216-632-1631
Ed-U-Cards	Games	New York, 1940s-1960s, manufacturer of children's card games		
Effanbee Doll Co.	Dolls		1026 W. Elizabeth Ave., Linden, NJ 07036-0006	908-474-8000
Einson-Freeman Co. Inc.	Games and Puzzles	Long Island City, NY, 1930s-1940s		
Empire Industries	Miscellaneous	acquired Buddy L in 1995	5150 Linton Blvd., DelRay, FL 33434	407-498-4000
Erector		See Meccano		
Ertl Company	Die-Cast Vehicle Toys, Banks, Farm Toys	1945-present	Highways 136 & 20, P.O. Box 500, Dyersville, IA 52040-0500	319-875-5607
Fisher-Price	Pull Toys, Plastics	1930-present; division of Mattel	636 Girard Ave., East Aurora, NY 14052	716-687-3449
Flexible Flyer Co.	Sleds		100 Tubb Ave., P.O. Box 1296, West Point, MS 39773	601-494-4732
Galoob Toys, Lewis	Action Figures, Play Sets		500 Forbes Blvd. S., San Francisco, CA 94080	415-873-0680
Gilbert, A.C.	Erector Sets, Action Figures	1916-1960s, See Meccano-Erector		
Goodenough & Woglom	Games	New York, NY, 1930s, producer of religious games		
Gund	Plush	1898-present	1 Runyons Lane, P.O. Box H, Edison, NJ 08818	908-248-1500
Harett-Gilmar	Games and Puzzles	Far Rockaway, NY, 1950s		
Hartland Plastics / Steven Manuf.	Western/Sports Figures		224 E. Fourth St., Hermann, MO 65041	314-486-5494
Hasbro Toy Group	Miscellaneous	1940s-present, previously known as Hassenfeld Bros., includes Kenner, Milton Bradley, Parker Bros., Tonka, Playskool	1027 Newport Ave., Pawtucket, RI 02862-1059	401-727-5582

Company Name	Product	Further Information	Current Address	Phone
Hood, C.I.	Games and Puzzles	Lowell, MA, 1870s-1920s		
Horsman Doll	Games and Dolls	New York, NY 1880-present, now division of Gatabox	200 Fifth Ave., New York, NY 10010	212-675-1720
Hubley Mfg. Co.	Vehicle Toys, Toy Guns, Banks	Lancaster, PA, 1892-1965, bought by Gabriel Industries in 1965		
Ideal Novelty & Toy Co.	Games and Dolls	Brooklyn, NY, 1907-1980s, now owned by Tyco Industries		
International Games	Games	Joliet, IL 1972-present, manufacturer of Uno card game; owned by Mattel	1 Uno Circle, Joliet, IL 60435	815-741-4000
Ives, Blakeslee & Williams	Miscellaneous	Bridgeport, CT, 1860s-1920s, later acquired by Lionel		
Ives, W. & S.B.	Games	1840s-1890s, sold to Parker Brothers in 1890s		
JA-RU Inc.	Miscellaneous		4030 Phillips Hwy., Jacksonville, FL 32207	904-733-9311
Jacmar Mfg. Co.	Games	Brooklyn, NY, 1950s, producer of electric educational games, a Marx subsidiary		
James Industries	Slinky	1945-present	P.O. Box 407, Hollidaysburg, PA 16648	814-695-5681
Jaymar Specialty / Jaymar	Miscellaneous	New York, NY 1930s-present, former Marx subsidiary	47 Scrantom St., Rochester, NY 14605-1054	716-454-7050
Justoys	Miscellaneous		50 West 23rd St., 7th Fl., New York, NY 10010-5205	212-645-6335
Kenner Products	G.I. Joe, Action Figures	division of Hasbro	615 Elsinore Pl., Cincinnati, OH 45202	513-579-4927
Kenton Lock Mfg.	Miscellaneous	Kenton, OH, 1894-1954, produced horse drawn wagons and pieces		
Kerk Guild	Games	Utica, NY, 1930s		
Kohner Bros.	Miscellaneous	New York, NY, 1950s		
LEGO Systems Inc.	Construction Toys, Interlocking Blocks		555 Taylor Rd., P.O. Box 1600, Enfield, CT 06083-1600	203-763-6731
Larami Corp.	Miscellaneous		340 N. 12th St., Philadelphia, PA 19107-1123	215-923-4900
Lionel Trains	Electric Toy Trains	1900-present	50625 Richard W. Blvd., Chesterfield, MI 48051	810-949-4100
Little Tikes	Preschool Toys	division of Rubbermaid	2180 Barlow Rd., Hudson, OH 44236	216-650-3000
Lowe, E.S., Co.	Games	New York, NY, 1920s-1973		
Lowe, Samuel	Games	Kenosha, WI, 1940s		
MPC	Model Kits	1960s-?		
Maisto International	Die-Cast Vehicle Toys		7751 Cherry Ave., Fontana, CA 92336	909-357-7988

Company Name	Product	Further Information	Current Address	Phone
Majorette / Solido	Vehicle Toys		1400 North West 93rd Ave., Miami, FL 33176	305-593-6016
Marklin	Toy Trains	distributor of German trains	16988 W. Victor Rd., P.O. Box 51319, New Berlin, WI 53151-0319	414-784-1095
Marks Brothers Co.	Games	Boston, MA, 1930s		
Marx, Louis / Marx Toy Company	Miscellaneous	1919-?, known primarily for play sets, figures, and tin; company has been sold many times and still produces some new items today	1130 N.W. 159th Dr., Miami, FL 33169	305-625-9000
Matchbox	Vehicle Toys	1947-present, now owned by Tyco Industries		
Mattel	Miscellaneous	1945-present, noted for Barbie and Hot Wheels; also produces Corgi vehicles in England; owns Fisher-Price; markets Cabbage Patch Kids and Wham-O's Hula Hoop and Frisbee	333 Continental Blvd., El Segundo, CA 90245-5012	310-252-2000
McFarlane Toys		Spawn and other action figures; company formerly known as Todd Toys	15155 Fogg St., Plymouth, MI 48170	313-414-3500
McLoughlin Brothers / John McLoughlin	Games	New York, NY, 1858-1920, prewar lithographed board games		
Meccano-Erector	Erector Sets, Construction Toys		1675 Broadway, 31st Fl., New York, NY 10019	212-397-0711
Mego	Action Figures	1954-1980s		
Milton Bradley	Games	1860-present, division of Hasbro	443 Shaker Rd. E., Longmeadow, MA 01028-3149	413-525-6411
Nintendo of America	Video Games		4820 150th Ave. NE, Redmond, WA 98052-5111	206-882-2040
Nylint	Vehicle Toys	1946-present	1800 Sixteenth Ave., Rockford, IL 61104-5491	815-397-2880
Ohio Art	Etch-a-Sketch, Vintage Tin Toys		One Toy Street, Bryan, OH 43506	419-636-3141
PEZ Candy	PEZ Candy Dispensers	founded in 1927 in Austria	35 Prindle Hill Rd., Orange, CT 06477	203-795-0531
Parker Brothers	Games	1880s to present, division of Hasbro	50 Dunham Rd., Beverly, MA 01915	617-927-7600
Playing Mantis	Johnny Lightning cars		3600 McGill St., Suite 300, P.O. Box 3688, South Bend, IN 46619-3688	219-232-0300
Playmates Toys	Action Figures		611 Anton Blvd. #600, Costa Mesa, CA 92626	714-428-2000
Playmobil USA	Figures, play sets		11-E Nichols Ct., Dayton, NJ 08810	908-274-0101

Company Name	Product	Further Information	Current Address	Phone
Playskool	Preschool Toys	division of Hasbro	1027 Newport Ave., Pawtucket, RI 02862-1059	
Pressman Toy	Games	1920s-present	200 Fifth Ave., Suite 1052, New York, NY 10010	212-675-7910
Radio Flyer	Wagons		6515 West Grand Ave., Chicago, IL 60635	800-621-7613
Reeves International	distributor of Breyer, Corgi		14 Industrial Rd., Pequannock, NJ 07440	201-694-5006
Remco Toys / Azrak-Hamway	Miscellaneous		1107 Broadway, Room 808, New York, NY 10010-2802	212-675-3427
Revell-Monogram	Model Kits		8601 Waukegan Rd., Morton Grove, IL 60053	708-966-3500
Russ Berrie & Co.	Plush		111 Bauer Dr., Oakland, NJ 07436	201-337-9000
Saalfield Publishing	Coloring Books, Puzzles, Games	Akron, OH, 1910-1970s		
Schaper	Games, Toys	Minneapolis, MN 1949-late 1980s		
Schoenhut	Wood Dolls	Philadelphia, PA		
Schuco / Lilliput Motor Company	Vehicle Toys		P.O. Box 447, Yerington, NV 89447	702-463-5181
Sega of America	Video Games		255 Shoreline Dr., Redwood City, CA 94065	415-508-2800
Selchow, E. G. / Selchow & Righter	Games	New York, NY, 1860s-1980s, sold to Coleco in 1986		
Smith-Miller	Vehicle Toys		P.O. Box 139, Canoga Park, CA 91305	818-703-8588
Solido		See Majorette		
Standard Toykraft	Games	Brooklyn, NY, 1940s-1960s		
Steiff USA	Teddy Bears, Plush	founded in 1880 in Germany	200 Fifth Ave., Suite 1205, New York, NY 10010	212-675-2727
Structo	Vehicle Toys	Freeport, IL, 1919-1975		
Tamiya America	Miscellaneous		2 Orion, Aliso Viejo, CA 92656-4200	714-362-2240
Thermos Co.	Lunch Kits		Rt. 75 East, Freeport, IL 61032	815-232-2111
Thinkway Toys	Miscellaneous; Toy Story		8885 Woodbine Ave., Markham, Ontario, Canada L3R 5G9	905-470-8883
Today's Kids		formerly Wolverine Supply and Mfg.	13630 Neutron Rd., Dallas, TX 75244	214-404-9335
Tomy America	Miscellaneous		450 Delta Ave., Brea, CA 92621	714-256-4990
Tonka	Vehicle Toys	1946-present, division of Hasbro		
Tootsietoy / Strombecker	Vehicle Toys, Other		600 N. Pulaski Rd., Chicago, IL 60624	312-638-1000
Topper Toys	Johnny Lightning, Dawn Dolls	1960s-1971		
Toy Biz	Action Figures		333 East 38th St., New York, NY 10016	212-682-4700
Transogram	Games	New York, 1915-1970s		

Company Name	Product	Further Information	Current Address	Phone
Trendmasters	Action Figures		611 North 10th St., Suite 555, St. Louis, MO 63101	314-231-2250
Tyco Industries	Miscellaneous	owns View-Master, Ideal, Matchbox	600 Midlantic Ave., Mt. Laurel, NJ 08054	609-234-7400
Uneeda Doll Co.	Dolls		200 Fifth Ave., Room 556, New York, NY 10010	212-675-3313
Unique Art	Miscellaneous	Newark, NJ, 1916-1952, bought by Louis Marx in 1952		
Williams, A.C.	Banks, Vehicle Toys	Chagrin Falls, OH, 1886-1938		
Winross	Die-Cast Trucks	1965-present	Box 23860, Rochester, NY 14692	716-381-5638
Wolverine	Miscellaneous	founded in Pittsburgh, PA in 1903; see Today's Kids		
Wyandotte		See All Metal Product Co.		

Toy Clubs Directory

Editor's Note: This is a partial list of toy collector clubs. To have your club listed free in the next edition of the Toy Shop Annual. *send pertinent information to: Editor, Toy Shop Annual, 700 E. State St., Iola, WI 54990.*

Action Canada
Specialty: Action Figures
Contact: Samantha Christopher
Address: P.O. Box 2810
Stony Plain, Alberta, Canada T7Z 1Y3
Phone: 403-892-2512

Action Figure Mania
Specialty: Pop, TV, Movie, & Music-Related Toys
Contact: Ralph Fearly
Address: 515 Ashbury St.,San Francisco, CA 94114
Newsletter: Yes
Members: 20 members

Alice in Wonderland Collectors Network
Specialty: Alice in Wonderland
Address: 2486 Brunswick Circle, Woodridge, IL 60517
Phone: 708-968-0664
Newsletter: Quarterly
Members: 40+ members

American Game Collectors Association
Specialty: 19th and 20th Century Games
Contact: AGCA Secretary
Address: 49 Brooks Ave., Lewiston, ME 04240
Annual Dues: $25
Members: 300 members

American International Matchbox
Specialty: Matchbox Cars
Address: 532 Chestnut St., Lynn, MA 01904
Phone: 617-595-4135
Annual Dues: $21
Newsletter: Monthly
Members: 800+ members

Antique Engine, Tractor & Toy Club
Address: 5731 Paradise Rd., Slatington, PA 18080
Phone: 215-767-4768
Annual Dues: $5
Newsletter: Quarterly
Members: 500 members

Antique Toy Collectors of America
Specialty: Antique Toys
Contact: Robert R. Grew
Address: 2 Wall St., 13th Floor, New York, NY 10005
Phone: 212-238-8803
Annual Dues: $50
Newsletter: Toychest; bi-annually
Members: 300 members

Aviation Toys
Specialty: Airplane Toys
Address: P.O. Box 845, Greenwich, CT 06836
Phone: 203-629-5270
Annual Dues: $35
Newsletter: The Plane News; quarterly
Members: 600+ members

Barbie Collectors of New England
Specialty: Barbie dolls and fashions
Contact: Dean Lillibridge
Address: 575 Pleasant St., Holyoke, MA 01040
Phone: 203-627-4973
Annual Dues: $25
Newsletter: Barbie O.N.E., quarterly
Members: 20-25, meetings every two months

Barbie Doll Collector Club International
Specialty: Barbie and Friends
Contact: Dora Lerch
Address: P.O. Box 586, N. White Plains, NY 10603
Annual Dues: $25
Newsletter: Barbie Newsletter International, bi-monthly
Members: 68 members

Batman TV Series Fan Club
Specialty: Batman
Address: P.O. Box 107, Venice, CA 90294
Phone: 310-841-5598
Annual Dues: $6
Newsletter: Quarterly
Members: 1,200 members

Big Little Book CLub
Specialty: Big Little Books
Contact: Larry Lowery
Address: P.O. Box 1242, Danville, CA 94526
Annual Dues: $12
Newsletter: The Big Little Times; bi-monthly
Members: 650+ members

Bobbin' Heads National Club
Specialty: Bobbin' Head Dolls
Contact: Barry Larkins
Address: P.O. Box 9297, Daytona Beach, FL 32120
Phone: 904-253-7040
Annual Dues: None

Canadian Toy Collectors Society
Specialty: All toys except dolls and trains
Address: 91 Rylander Blvd., Unit 7, Suite 245
Scarborough, Ontario, M1B 5M5 Canada
Phone: 416-287-0242
Annual Dues: $40
Newsletter: Yes
Members: 200+ members

Captain Action Collectors Club
Specialty: Captain Action
Address: P.O. Box 2095, Halesite, NY 11743
Phone: 516-423-1801
Newsletter: Send SASE

Care-a-Lot: Care Bear Collectors' Club
Specialty: Care Bears
Contact: Chris McClure
Address: Rt. 3 Box 62-B, Winfield, KS 67156
Phone: 316-221-0078
Annual Dues: $5
Newsletter: Heartline, bi-monthly
Members: 40 members

Cascade Classic Cycle Club
Address: 7935 SE Market St., Portland, OR 97215
Phone: 503-775-2688
Annual Dues: $15
Newsletter: Yes
Members: 250 members

Chatty Cathy Collectors Club
Specialty: Mattel Chatty Cathy Dolls
Contact: Lisa Eisenstein
Address: 2610 Dover St., Piscataway, NJ 08854
Annual Dues: $25
Newsletter: Chatty News; quarterly
Members: 200 members

Classic Action Figure Collectors Club
Specialty: Action Figures 1960s present
Contact: Diana Bonavita
Address: P.O. Box 2095, Halesite, NY 11743
Annual Dues: $24
Newsletter: Collect 'Em All; quarterly

Classic Bicycle & Whizzer Club
Specialty: Bicycles
Address: 35892 Parkdale, Livonia, MI 48150
Annual Dues: $18
Newsletter: Monthly
Members: 500 members

Coca-Cola Collectors Club
Specialty: Coca-Cola
Contact: Bill Combs
Address: 400 Montemar Ave., Baltimore, MD 21228-5213
Phone: 410-744-0816

Cracker Jack Collectors Association
Specialty: Cracker Jack toys
Contact: Ron Toth
Address: 72 Charles St., Rochester, NH 03867
Phone: 603-355-2062
Annual Dues: $18 or $24
Newsletter: Prize Insider; monthly
Members: 150 members

Dagwood & Blondie Fan Club
Specialty: Dagwood & Blondie
Contact: Rod Carnahan
Address: 541 El Paso St., Jacksonville, TX 75766
Phone: 903-586-1355
Annual Dues: None
Newsletter: None
Members: 10 members

Dark Sentinel/Batman
Specialty: Batman
Contact: Carlos R. Hernandez
Address: 111 Tallavana Dr., Havana, FL 32333
Annual Dues: $40
Newsletter: Quarterly
Members: 63 members, $55 foreign dues

Dark Shadows Fan Club
Specialty: Dark Shadows
Contact: Louis Wendruck
Address: P.O. Box 69A04, West Hollywood, CA 90069
Annual Dues: $30
Newsletter: Yes

Dick Tracy Fan Club
Specialty: Dick Tracy
Contact: Larry Doucet
Address: P.O. Box 632, Manitou Springs, CO 80829-0632
Newsletter: 5 times/year
Members: 1,500 members

Dinky Toy Club of America
Specialty: Dinky Toys
Contact: Jerry Fralick
Address: P.O. Box 11, Highland, MD 20777
Phone: 301-854-2217
Annual Dues: $20 US
Newsletter: DTCA Newsletter
Members: 270 members

Dinosaur Toy & Model Collecting Club
Contact: Mike Fredericks
Address: 145 Bayline Circle, Folsom, CA 95630
Annual Dues: $19
Newsletter: Prehistoric Times; bimonthly

Disneyana Dreamers of San Diego County
Specialty: Disneyana
Address: P.O. Box 106, Escondido, CA 92033
Phone: 619-747-2990
Annual Dues: $12
Newsletter: Monthly

Dukes of Hazzard Fan Club
Specialty: TV Character Club
Contact: Michael Streit
Address: 1011 N. Lake St., Aurora, IL 60506-2415
Annual Dues: $12
Newsletter: Quarterly
Members: 200 members

Durham Classics Collectors Club
Specialty: Durham Diecast
Contact: Gary A. Fry
Address: 24 Montrose Crescent, Unionville, Ontario, Canada L3R 7Z7
Annual Dues: $27
Newsletter: Quarterly
Members: 200+ members

East Koast Kiddle Klub
Specialty: Liddle Kiddles
Contact: Cindy Sabulis
Address: 3 Stowe Dr., Shelton, CT 06484
Phone: 203-926-0176
Members: 12 members

Emerald City Club
Specialty: Wizard of Oz and Judy Garland
Contact: Lee Jenkins
Address: 153 E. Main St., New Albany, IN 47150
Annual Dues: $10
Newsletter: Quarterly
Members: 141 members

Emergency Vehicle Collectors Club
Specialty: Miniature police, fire & EMS vehicles
Contact: Christopher Barlau
Address: P.O. Box 14616, Minneapolis, MN 55414
Phone: 612-872-8627
Annual Dues: $20
Newsletter: EVCC Newsletter; bimonthly
Members: 200+ members

Ertl Replica Club
Specialty: Ertl Toys
Contact: Karen Sands
Address: P.O. Box 500, Dyersville, IA 52040
Phone: 319-875-2000
Annual Dues: $8
Newsletter: Bi-monthly
Members: 40,000 members

Fanimaniacs
Specialty: Warner Bros. Animaniacs
Contact: Michael Macomber
Address: #223 Linden Hill Apts., Lindenwold, NJ 08021
Newsletter: Fanimaniacs Newsletter

Fast Food Toy Club of Northern Illinois
Specialty: Restaurant Premiums
Contact: Jeff Escue
Address: 164 Larchmont Ln., Bloomingdale, IL 60108
Phone: 708-307-7320
Annual Dues: None
Newsletter: None
Members: 18 members

Fisher-Price Collectors Club
Contact: Jeanne Kennedy
Address: 1442 N. Ogden, Mesa, AZ 85205
Annual Dues: $15
Newsletter: Quarterly
Members: 250 members

Friends of Hopalong Cassidy
Specialty: Hopalong Cassidy
Contact: John Spencer
Address: 4613 Araby Church Rd., Frederick, MD 21705
Annual Dues: $15
Newsletter: Quarterly
Members: 300 members

Friends of Hoppy Club
Specialty: Hopalong Cassidy
Contact: Laura Bates
Address: 6310 Friendship Dr., New Concord, OH 43762-9708
Phone: 614-826-4850
Newsletter: Hoppy Talk

Friends Toy Collectors Association
Specialty: Toys/Dolls
Contact: Glenda Hodson
Address: 10227 Atkins Rd., Bentonville, AR 72712
Annual Dues: $14
Newsletter: None
Members: 14 members

Fuddsters NW
Specialty: Elmer Fudd
Contact: Monte Johnson
Address: One SW Columbia #1610, Portland, OR 97258
Phone: 800-224-0158

G.I. Joe Collectors Club
Specialty: G.I. Joe
Contact: James DeSimone
Address: 150 S. Glenoaks Blvd., Burbank, CA 91510
Phone: 818-563-1179
Annual Dues: $20
Newsletter: Monthly
Members: 1,468 members

G.I. Joe Collectors Club
Specialty: G.I. Joe
Address: 12513 Birchfalls Dr., Raleigh, NC 27614
Phone: 919-847-5263
Annual Dues: $35

Gamers Alliance
Specialty: Games
Contact: H.M. Levy
Address: P.O. Box 197-TSA, East Meadow, NY 11554
Annual Dues: $25-30
Newsletter: Quarterly

Garfield Collectors Society
Specialty: Garfield collectibles
Contact: David L. Abrams
Address: 7744 Foster Ridge Rd., Germantown, TN 38138
Annual Dues: $15
Newsletter: Five issues
Members: 100 members

Garfield Connection, The
Specialty: Garfield
Contact: Denise Karl
Address: 2 Lyons Rd., Armonk, NY 10504
Phone: 914-273-3575
Annual Dues: $9
Newsletter: Monthly
Members: 80 members

Gateway Farm Toy Club
Specialty: Farm Toys
Address: 5583 Jimtown Rd., Oakdale, IL 62268
Phone: 618-566-2166
Annual Dues: $10
Newsletter: Yes
Members: 60 members

Gilligan's Island Fan Club
Specialty: Gilligan's Island
Contact: Bob Rankin
Address: P.O. Box 25311, Salt Lake City, UT 84125
Phone: 801-275-5729
Annual Dues: $15
Newsletter: Island News; quarterly
Members: 800 members

Girder and Panel Collectors Club
Specialty: Kenner Girder and Panel Sets
Contact: Ed Sterling
Address: Box 494, Bolton, MA 01740
Phone: 508-779-6058
Annual Dues: $10
Newsletter: Quarterly
Members: 90 members

Good Bears of the World
Specialty: Teddy Bears for Abused Children
Contact: Terri Stong
Address: P.O. Box 13097, Toledo, OH 43613
Phone: 419-531-5365
Annual Dues: $12
Newsletter: Quarterly
Members: 4,000 members

Herbie Love Bug Lovers
Specialty: Anything related to #53 Herbie the Love Bug
Contact: Melissa Jess
Address: 3121 E. Yucca St., Phoeniz, AZ 85028
Annual Dues: Postage
Newsletter: Yes
Members: 27 Members

Holly Hobbie Collectibles of America
Specialty: Holly Hobbie
Contact: Helen McCale
Address: P.O. Box 397, Butler, MO 64730-0397
Phone: 816-679-3690
Annual Dues: $25
Newsletter: Bi-monthly

Hot Wheels Collectors Club, Tampa Bay Chapter
Specialty: Hot Wheels
Contact: Joe Madonia
Address: 6735 Hammock Rd. #105, Port Richey, FL 34668
Annual Dues: $10
Newsletter: Collector Club Newsletter; monthly
Members: 48 members

Howdy Doody Memorabilia Collectors Club
Specialty: Howdy Doody
Address: 8 Hunt Ct., Flemington, NJ 08822
Phone: 908-782-1159
Annual Dues: $18
Newsletter: Monthly
Members: 200 members

Ideal Doll & Toy Collectors Club
Specialty: Ideal Dolls/Toys
Contact: Judith Izen
Address: P.O. Box 623, Lexington, MA 02173
Phone: 617-862-2994
Annual Dues: $10
Newsletter: Ideal Doll News

Imagination Guild
Specialty: Disneyana
Address: P.O. Box 907, Boulder, CA 95006
Phone: 408-335-2755
Annual Dues: $32
Newsletter: Monthly

International Big Jim Collectors' Association
Specialty: Mattel's Big Jim
Contact: Kip North
Address: 904 Woodland Dr., Wheeling, IL 60090
Phone: 847-465-8040
Annual Dues: $5
Newsletter: Quarterly

International Diecast Model Collectors
Contact: Bob Patterson
Address: 5882 Fulton Dr. NW #210, Canton, OH 44718
Annual Dues: $18
Newsletter: Quarterly
Members: 65 members

International Figure Kit Club
Specialty: Figure Model Kits
Contact: Gordy Dutt
Address: P.O. Box 201, Sharon Center, OH 44274
Phone: 216-239-1657
Annual Dues: $25
Newsletter: Quarterly

Johnny Lightning News Flash
Specialty: Johnny Lightning Cars
Contact: Playing Mantis Toys
Address: 3600 McGill St., Suite 300, P.O. Box 3688, South Bend, IN 46619-3688
Phone: 219-232-0300
Members: 5,000 members

Larry's Traders
Specialty: Trading Cards
Address: 534 Gates, Aurora, IL 60505
Phone: 708-851-0199

Lionel Collectors Club of America
Specialty: Lionel Trains
Address: P.O. Box 479, LaSalle, IL 61301
Phone: 815-654-1705
Annual Dues: $30
Newsletter: Bi-monthly

Lionel Operators Train Society
Specialty: Lionel Trains
Address: P.O. Box 62240, Cincinnati, OH 45241
Phone: 513-874-5583
Annual Dues: $28
Newsletter: Bi-monthly
Members: 2,300 members

Lost in Space Fan Club
Specialty: Lost in Space
Address: 550 Trinity A Club, Westfield, NJ 07090
Phone: 908-789-7323
Annual Dues: None
Newsletter: Yes
Members: 50 members

Man from U.N.C.L.E. Collectors Club
Specialty: Man from U.N.C.L.E.
Contact: Jon Heitland
Address: 1922 Timberedge, Iowa Falls, IA 50126
Annual Dues: None
Newsletter: None
Members: 10 members

Marble Collectors Society of America
Contact: Stanley A. Block
Address: P.O. Box 222, Trumbull, CT 06611
Annual Dues: $12
Newsletter: Marble Mania; quarterly
Members: 1,500 members

Matchbox Collectors Club
Specialty: Matchbox Cars
Address: P.O. Box 977, Newfield, NJ 08344
Phone: 609-697-2800
Annual Dues: $14
Newsletter: Yes
Members: 1,500 members

Matchbox Int'l Collectors Association (MICA)
Specialty: Matchbox Cars
Contact: Rita Schneider
Address: 574 Canewood Crescent, Waterloo, Ontario, Canada N2L 5P6
Phone: 519-885-0529
Annual Dues: $30
Newsletter: Bi-monthly
Members: 6,000 members, $35 dues in Canada

Matchbox U.S.A.
Specialty: Matchbox Cars
Contact: Charles Mack
Address: 62 Saw Mill Rd., Durham, CT 06422
Phone: 203-349-1655
Annual Dues: $25
Newsletter: Monthly
Members: 1,600 members, $28 overseas dues

McDonald's Collectors Club
Specialty: McDonald's Memorabilia
Contact: Linda Gegorski
Address: 424 White Rd., Fremont, OH 43420
Annual Dues: $20
Newsletter: Quarterly; McDonald's Collectors Club News
Members: 1,200 members

McDonald's Collectors Club Sunshine Chapter
Specialty: McDonald's Toys
Contact: Bill & Pat Poe
Address: 220 Dominica Circle E., Niceville, FL 32578
Phone: 904-897-2606
Annual Dues: $10
Newsletter: Sunshine Express; monthly
Members: 90+ members

McDonald's Collectors' Club (Northeast Ohio Chapter)
Specialty: McDonald's Toys
Contact: Ken Brady
Address: 3381 W. 130 St., Cleveland, OH 44111
Phone: 216-941-7127

Mechanical Bank Collectors of America
Specialty: Mechanical Banks
Contact: H.E. Muhlheim
Address: P.O. Box 128, Allegan, MI 49010
Phone: 616-673-4509
Annual Dues: $30
Newsletter: Quarterly
Members: 370 members

Miniature Piano Enthusiast Club
Address: 5815 N. Sheridan Rd. Ste 202, Chicago, IL 60660
Phone: 312-271-2970
Annual Dues: $8
Newsletter: Quarterly

Mountain Ears Disneyana Club
Specialty: Disneyana Fan Club
Contact: Karen Eagle
Address: 4803 S. 164th St., Seattle, WA 98188
Annual Dues: $12
Newsletter: Mt. Ears News; 6X year
Members: 70 members

Mouse Club, The
Specialty: Disneyana
Contact: Kim and Julie McEuen
Address: 2056 Cirone Way, San Jose, CA 95124

Munsters & Addams Family Fan Club
Specialty: Munsters/Addams Family
Contact: Louis Wendruck
Address: P.O. Box 69A04, West Hollywood, CA 90069

National Fantasy Fan Club for Disneyana
Specialty: Disneyana
Address: P.O. Box 19212, Irvine, CA 92713
Phone: 714-731-4705
Annual Dues: $20
Newsletter: Monthly
Members: 2,000 members

National Model Railroad Association
Specialty: Model Railroading
Address: 4121 Cromwell, Chattanooga, TN 37421
Phone: 615-892-2846
Annual Dues: $24
Newsletter: Quarterly
Members: 26,000 members

New Switzerland Model Railroad Club
Specialty: Model Railroading
Address: 12226 Albrecht Rd., Alhambra, IL 62001
Phone: 618-654-3955
Annual Dues: $10
Newsletter: Monthly
Members: 35 members

North American Diecast Toy Collectors Association
Specialty: Diecast Toys
Contact: Dana Johnson
Address: P.O. Box 1824, Bend, OR 97709-1824
Annual Dues: $15
Newsletter: Diecast Toy Collector; monthly
Members: 80 members

Northeast Ohio Star Wars Collecting Club
Specialty: Star Wars
Contact: Chris Fawcett
Address: 426 Washburn Rd., Tallmadge, OH 44278
Phone: 330-633-0295

Northeast Toy Soldier Society
Specialty: Military & Civilian Figures
Contact: Arley Pett
Address: 12 Beach Rd, Gloucester, MA 01930
Phone: 508-283-2612
Annual Dues: $10.00
Newsletter: Monthly mailings and meetings
Members: 75 members

Northland Hot Wheels Club
Specialty: Hot Wheels
Contact: Paul Dyke
Address: 661 Bridle Ridge Rd., Eagan, MN 55123
Phone: 612-452-6560
Annual Dues: $2
Newsletter: None
Members: 43 members

On Line with Betsy McCall
Specialty: Betsy McCall Dolls
Contact: Marci Van Ausdall
Address: P.O. Box 946, Quincy, CA 95971
Phone: 916-283-2770
Annual Dues: $10
Newsletter: Quarterly

Original Paper Doll Artist Guild
Specialty: Paper Dolls
Address: P.O. Box 14, Kingsfield, ME 04947
Phone: 207-165-2500
Annual Dues: $15
Newsletter: Quarterly
Members: 350 members

Peanuts Collector Club
Specialty: Peanuts Character Toys
Contact: Andrea Podley
Address: 539 Sudden Valley, Bellingham, WA 98226
Phone: 206-733-5209
Annual Dues: $16
Newsletter: Yes
Members: 700 members

Popeye Fan Club
Contact: Mike and Debbie Brooks
Address: 1001 State St., Chester, IL 62233
Phone: 618-826-4567
Annual Dues: $8
Newsletter: Quarterly

Post Car Registry
Specialty: Post Cereal Premium Cars
Contact: Tom Morgan
Address: 812 North Third St., St. Peter, MN 56082
Phone: 507-931-4369
Annual Dues: $5
Newsletter: Quarterly
Members: 50 members

PVC Collectors Club
Specialty: PVC figures
Contact: Colleen Lewis
Address: 10120 Main St., Clarence, NY 14031
Phone: 716-759-7451
Annual Dues: $24
Newsletter: PVC Collector; quarterly

Roy Rogers & Dale Evans Collectors
Contact: Nancy Horsely
Address: P.O. Box 1166, Portsmouth, OH 45662
Phone: 614-353-0900
Annual Dues: $15
Newsletter: Quarterly
Members: 1,000 members

Schoenhut Collector's Club
Specialty: Schoenhut Toys
Address: 45 Louis Ave., West Seneca, NY 14224
Annual Dues: $15
Newsletter: Quarterly
Members: 400 members

Schoenhut Toy Collectors
Specialty: Schoenhut
Address: 1916 Cleveland St., Evanston, IL 60202
Phone: 708-866-7165
Annual Dues: $10
Newsletter: Yes
Members: 350+ members

Smurf Collectors' Club International
Contact: Suzanne Lipschitz
Address: 24 Cabot Rd. W, Massapequa, NY 11758
Annual Dues: SASE
Newsletter: Quarterly
Members: 2,600 members

Smurfing Times
Address: P.O. Box 132, Holmes, PA 19043
Phone: 610-876-7779
Newsletter: Bi-monthly
Members: 50 members

SNEW
Specialty: Tomy Pocket Cars
Contact: Bob Blum
Address: 8 Leto Rd., Albany, NY 12203
Annual Dues: $12
Newsletter: SNEW, four times/year
Members: 65 members

Southern California Meccano and Erector Club
Specialty: Construction Toys
Contact: Joel Perlin
Address: P.O. Box 7653, Porter Ranch Station, Northridge, CA 91327
Phone: 805-985-5498
Annual Dues: $12
Newsletter: Quarterly
Members: 250 members

Spect-tacular News
Specialty: Spec Cast
Address: P.O. Box 368, Dyersville, IA 52040
Phone: 319-875-8706
Annual Dues: $8
Newsletter: Yes

Star Trek: The Official Fan Club
Address: P.O. Box 111000, Aurora, CO 80011
Phone: 800-878-3326
Annual Dues: $11.95
Newsletter: Yes
Members: 10,000+ members

Star Wars Collectors Club
Specialty: Star Wars Collectibles
Contact: David W. Carr
Address: 20201 Burnt Tree Lane, Walnut, CA 91789
Annual Dues: $5
Newsletter: Yes

Steiff Collectors Club
Address: P.O. Box 798, Holland, OH 43528
Phone: 419-865-3899
Annual Dues: $8
Newsletter: Quarterly
Members: 3,000 members

Still Bank Collectors Club of America
Specialty: Still Banks
Address: 1456 Carson Ct., Homewood, IL 60430
Phone: 708-799-1732
Annual Dues: $35
Newsletter: Yes
Members: 400+ members

Tampa Bay FL Area Chapter Hot Wheels Collector Club
Specialty: Hot Wheels/Johnny Lightning
Contact: Oscar and Kitty Vaughn
Address: 4531 W. Lambright, Tampa, FL 33614
Annual Dues: $10.00
Newsletter: Yes/Monthly
Members: 70 members

Three Stooges Collectors Club
Specialty: Three Stooges
Address: P.O. Box 72, Skokie, IL 60076
Phone: 708-432-9270
Annual Dues: None

Toonerville Trolley Collectors
Contact: Asa Sparks
Address: 6045 Camelot Ct., Montgomery, AL 36117
Phone: 205-27-0687
Annual Dues: $10
Newsletter: Yes
Members: 90 members

Toy Car Collectors Club
Specialty: Vehicle Toys
Contact: Peter Foss
Address: 33290 W. 14 Mile Rd. #454, West Bloomfield, MI 48322
Phone: 810-682-0272
Annual Dues: $15
Newsletter: Toy Car Magazine; quarterly

Toy Gun Collectors of America
Address: 312 Starling Way, Anaheim, CA 92807
Phone: 714-998-9615
Annual Dues: $15
Newsletter: Quarterly
Members: 300 members

Train Collectors Association
Specialty: Toy Trains
Contact: Tony D'Alessandro
Address: 300 Paradise Ln., Strasburg, PA 17579
Annual Dues: $20
Newsletter: National Headquarters News
Members: 26,000 members

Troll Monthly
Specialty: Trolls
Address: 5858 Washington St., Whitman, MA 02382
Phone: 800-85-TROLL
Annual Dues: $24
Newsletter: Monthly
Members: 300+ members

Tucson Miniature Auto Club
Specialty: Toy Vehicles
Contact: Lou Pariseau
Address: 1111 E. Limberlost Dr. #164, Tucson, AZ 85719
Annual Dues: $10
Newsletter: Mini Auto Gazette; monthly
Members: 80 members

Turtle River Toys
Specialty: Farm Toys
Address: Rt. 1, Box 44, Manvel, ND 58256
Phone: 701-699-3577
Annual Dues: $16
Newsletter: Monthly

We Love Lucy: The International Lucille Ball Fan Club
Specialty: Lucille Ball
Contact: Thomas Watson
Address: P.O. Box 56234, Sherman Oaks, CA 91413
Annual Dues: $20
Newsletter: Quarterly
Members: 800 members

West Phoenix Doll Club
Address: 225 E. Adams, Phoenix, AZ 85051
Phone: 602-841-0761

White Knob Windups
Specialty: Plastic wind-up toys
Address: 61 Garrow St., Auburn, NY 13021
Phone: 315-253-9131
Annual Dues: $24
Newsletter: Monthly
Members: 70 members

Winross Collectors Club of America
Specialty: Winross trucks
Contact: Rhoda Groff
Address: P.O. Box 444, Mt. Joy, PA 17552-0444
Phone: 717-653-7327
Annual Dues: $15
Newsletter: Monthly
Members: 4,200 members

Action Figures

Action Figures Emporium
3524 Mont Martre Dr.
Orlando, FL 32822
407-273-9491

Action Figures from the 80s
1400 Golden #109
Golden, CO 80401
303-384-9735

Action Images
101 Middlesex Tpke.
Burlington, MA 01803
508-663-2001

Adventure Into Comics & Collectables
1033 E. Semoran Blvd.
Casselberry, FL
407-830-9636

Al Chin
39 Dorchester Dr.
East Brunswick, NJ 08816
908-613-0907

Betsy Megalos
105 McCoy Ct
Cary, NC 27511
919-481-1931

Brian Semling
W730 State Rd. 35/54
Fountain City, WI 54629
608-687-7572

Cosmic Toys / Zane Coopersmith
6410 Elliot Place
Hyattsville, MD 20783
301-559-2154

Darkside Toys
245 South 350 East
Angola, IN 46703-9292
219-665-1065

Fantasy Toys
P.O. Box 365
Cornish Flat, NH 03746
603-542-2378

Figures
P.O. Box 19482
Johnston, RI 02919
401-946-5720

Heroes Unlimited
P.O. Box 453
Oradell, NJ 07649
201-385-8815

John Wesley Hardin
1046 Tasker St.
Philadelphia, PA 19148
610-446-6513

National Collector
6113 Clark Center Ave.
Sarasota, FL 34238
941-927-8398

Puzzle Zoo
1413 3rd St. Promenade
Santa Monica, CA 90401
310-260-1547

Quakerhead Collectibles
P.O. Box 255
Broomall, PA 19008
215-747-6077

Rebel Toy
2300 California St.
Placentia, CA 92870
714-579-0673

Rich Intihar
7552 Saint Clair Ave. #B
Mentor, OH 44060
216-352-8190

Samantha and Bruce Christopher
P.O. Box 2810
Stony Plain, Alberta, Canada T7Z 1Y3
403-892-2512

Sanart Collectibles
482 81st St.
Brooklyn, NY 11209
718-680-4242

Serpa's Sportscards
4301 W. Princeton Ave.
Fresno, CA 93722
209-276-0730

Splash Page Comics
1007 E. Patterson St.
Kirksville, MO 63501
816-665-7623

Stand Up Comics
2402 University Ave. W.
Saint Paul, MN 55114
612-426-6427

Star Wars and Stuff
P.O. Box 24344
Fort Worth, TX 76124
817-536-8990

Vince Jones
P.O. Box 172
Davenport, IA 52805

Announcements

Antique World Shows
P.O. Box 34509
Chicago, IL 60634
847-526-1645

Atlantic Group
P.O.Box 217
Swansea, MA 02777
508-675-8745

Big-D Super Collectibles Show
P.O. Box 200725
Arlington, TX 76006
817-261-8745

Bob Smith
62 West Ave.
Fairport, NY 14450
716-377-8394

Bob Valdez
880 E. 1050 N.
Ogden, UT 84404
801-782-1124

Bruce Beimers
1720 Rupert St. NE
Grand Rapids, MI 49505
616-361-9887

Carolina Hobby Expo
3452 Odell School Rd.
Concord, NC 28027
704-786-8373

Creative World Associates
11910 Asbury Dr.
Fort Washington, MD 20744
301-292-5409

Don DeSalle
5106 Knollwood Ln.
Anderson, IN 46011
317-649-8971

George Zaninovich
310-832-2282

JD Productions
P.O. Box 726
Cherry Hill, NJ 08003
609-795-0436

John Carlisle
409 Bewley Bldg.
Lockport, NY 14094

K & J Entertainment
P.O. Box 453
Oradell, NJ 07649
201-385-8815

Leading Edge Productions
P.O. Box 31388
Palm Beach Gardens, FL 33420
407-694-7982

Majic Productions
948 Semoran Blvd.
Casselberry, FL 32707
407-260-8869

Paul DeCarlo
P.O. Box 1354
North Massapequa, NY 11758
516-289-7398

R & S Enterprises
34 N. Vintage Rd.
Paradise, PA 17562
717-442-4279

Scott Antique Markets
P.O. Box 60
Bremen, OH 43107
614-569-4112

Skip's Toy & Doll Show
P.O. Box 88266
Carol Stream, IL 60188
708-682-8792

Southeastern Collectible Expo
1819 Peeler Rd.
Vale, NC 28168
704-276-1670

Toys N' Stuf
P.O. Box 2037
San Bernadino, CA 92406
909-880-8558

Wayne Gateley Productions
P.O. Box 221052
Sacramento, Ca 95822
916-448-2655

Zurko's Midwest Promotions
211 W. Green Bay St.
Shawano, WI 54166
715-526-9769

Antique Toys, For Sale

Art & Toys
P.O. Box 201/Nordwall F4
477G8 Krefeld, Germany

Ben De Voto
756 Craig Dr.
Kirkwood, MO 63122
314-821-8588

Bill Campbell
1221 Little Brook Ln.
Birmingham, AL 35235
205-853-8227

Classic Tin Toy
P.O. Box 193
Sheboygan, WI 53082
414-693-3371

Dan Curran
223 E. Oak Ave.
Moorestown, NJ 08057
609-222-1005

Darrow's Fun Antiques
1101 1st Ave.
New York, NY 10021
212-838-0730

Lily Pad Antiques and Old Toys
756 Broadway
Tacoma, WA 98402
206-627-6858

N.J. Nostalgia Hobby
401 Park Ave.
Scotch Plains, NJ 07076
908-322-2676

Norman's Olde Store
126 W. Main St.
Washington, NC 27889
919-946-3448

Olde Tyme Toy Shop
120 S. Main St.
Fairmount, IN 46928
317-948-3150

Olde Tyme Toys
209 W. Main St.
Richmond, KY 40475
606-623-8832

Paul and Jean Stage
7558 Preakness Dr.
Caledonia, IL 61011
303-225-7980

Plymouth Rock Toy Co.
P.O. Box 1202
Plymouth, MA 02362
508-746-2842

Richard Johnson
P.O. Box 27093
Prescott Valley, AZ 86312
502-775-4714

Shelby Messinger
9 Winthrop Rd.
Jamesburg, NJ 08831
609-395-7604

Toys and More
P.O. Box 3192
Oshkosh, WI 54903
414-231-0185

Antique Toys, Wanted

Frank A. Najbart
2736 Bee Tree Ln.
St. Louis, MO 63129-5610
314-846-2444

Taylor's Toys
6720 Quiet Lane
Brentwood, TN 37027
615-776-3104

Auctions

Barry Goodman
P.O. Box 218
Woodbury, NY 11797
516-338-2701

Charles Rastelli
122 S. 13th St.
Weirton, WV 26062
304-748-2303

Fontaine's Auction Gallery
173 S. Mountain Rd.
Pittsfield, MA 01201
413-448-8922

Joel Magee
19 Edgewater Ln.
Dakota Dunes, SD 57049
712-277-8388

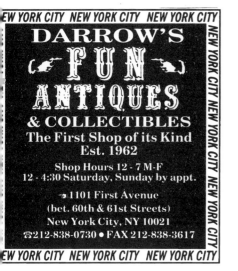

Larry Martin Toy Sales
P.O. Box 333
Clinton, IL 61727
217-935-3245

Banks: Die-Cast and Plastic

Ar-Jay Sales Co.
715 Mt. Pleasant
Houghton Lake, MI 48629
517-422-5972

Bruce Johnson Toy Talk
702 Steeplechase Rd.
Landisville, PA 17538

D & D Toys
8724 Earlmar Ct.
Elk Grove, CA 95624
916-685-8323

Don's Collectable Toys
506 Wilson Dr.
Hockessin, DE 19707
302-239-5137

Evers Toy Store
204 1st Ave. East
Dyersville, IA 52040
319-875-2438

Nickel's Amoco
8207 Liberty Rd.
Baltimore, MD 21244
410-655-9606

Small Wheels of America
28 Orchard Rd.
Farmington, CT 06032
203-521-0325

Toys Plus
2353 N. Wilson Way
Stockton, CA 95205
209-944-9977

Books on Toys

Blystone's
2132 Delaware Ave.
Pittsburgh, PA 15218
412-371-3511

Buses & Special Vehicles, For Sale

Streamline Toys
15009 Chestnut Dr.
Burnsville, MN 55306
612-898-3236

Buses & Special Vehicles, Wanted

Fin's Deckers & Trams
8215 4th Ave.
Brooklyn NY 11209
718-833-0109

Buttons, Pinback, Pins

Oasis of Quality
336 Shamrock Rd.
St. Augustine, FL 32086
904-826-0835

Cars, For Sale

Adkins Collectible Toys, Ltd.
1913 W. Puetz Rd.
Oak Creek, WI 53154
414-761-1020

Asheville Diecast
1412 Brevard Rd.
Asheville, NC 28806
704-667-9690

Auto Toys
P.O. Box 81385
Bakersfield, CA 93380
805-588-2277

Barker's Collectible Toys
202 S. Front St.
Dowagiac, MI 49047
616-782-4125

Bob & Di-Versions
P.O. Box 2115
Broken Arrow, OK 74013
918-455-5037

Budget-Minder Collectibles
701 E. Bay St.
Charleston, SC 29403
803-577-7695

C & D Collectible Diecast
P.O. Box 66008
Mobile, AL 36660
334-476-4291

C.A.R.S.
R.R. 3, Box 71W
Hermann, MO 65041
573-252-4398

Champion Toys & Automobilia
R.R. 1, Box 1858
Kennebunkport, ME 04046
207-985-2292

Charles & Sons, Ltd.
24 Woodbine Ave.
Northport, NY 11768
516-261-1523

Coast to Coast Toys
165 Province Line Rd.
Wrightstown, NJ 08562
609-758-1486

D & K Toy Collectibles
8112 Fulton St. E.
Ada, MI 49301
616-676-8876

Dan Wells Antique Toys
7311 Highway 329, Suite 601
Crestwood, KY
502-896-0740

Dual Connection
Box 569
Gibsonia, PA 15044
412-381-1143

Evergreen Import
2525 Royal Ln., Stuie 327
Dallas, TX 75229
214-488-8138

EWA & Miniature Cars USA
369 Springfield Ave.
Berkeley Heights, NJ 07922
908-665-7811

Ifitzgot Wheels
104 West Virginia
McKinney, TX 75069
972-548-9550

Louis Pan Use
5627 47th Ave. SW
Seattle, WA 98136
208-932-9787

Michael Silver
509 Danielle Ct.
Roseville, CA 95747
916-782-4800

Mike Silver
916-331-6527

Model Cars & Trains Unlimited
28 Arthur Ave.
Blue Point, NY 11715
516-363-2134

Paul Garnand Sales
203 Campus Dr., P.O. Box 68
Garden City, KS 67846
316-275-9552

People Kars
290 Third Avenue Extension
Rensselaer, NY 12144
518-465-0477

Pete's Miniature Motors
406 Main St.
Spring, TX 77373
713-288-1236

R & B Collectibles
P.O. Box 406
Frenchtown, NJ 08825
908-996-2141

Roger Wills Toys and Collectibles
520 E. 78th Lane
Merrillville, IN 46410
219-769-5117

San Remo Hobby
93 Beaver Dr.
Kings Park, NY 11754
516-724-5722

Slot Car Johnnie's
1195 Crestview St.
Reynoldsburg, OH 43068
614-864-8538

Cars, Wanted

Larry S. Angian, Jr.
4 Eastern Ave.
Watervliet, NY 12189
518-274-0247

VW Toy Collectors c/o Jess
3121 E. Yucca St.
Phoeniz, AZ 85028
602-867-7672

Cereal Boxes & Premiums

Chris Serratore
218 N. Main St.
Amber, PA 19002
215-628-3213

Flake World / Scott Bruce
P.O. Box 481
Cambridge, MA 02140
617-492-5004

Guy Himber
15239 Rayneta Dr.
Sherman Oaks, CA 91403

Roland Coover
1537 E. Strasburg Rd.
West Chester, PA 19380
610-692-3112

Character Toys, For Sale

5 & Dime
40 N. Union St.
Lambertville, NJ 08530
609-397-4957

52 Girls Collectibles
P.O. Box 36
Morral, OH 43337
614-465-6062

Absolute Antiques
P.O. Box 5199
Woodbridge, VA 22194
703-670-5603

Adrienne Warren
1032 Feather Bed Lane
Edison, NJ 08820
908-381-7083

America's Heroes
302 Crain Hwy SW, Ste 4
Glen Burnie, MD 21061
410-768-7895

Animation Plus
537 W. Diversey Pky.
Chicago, IL 60614
773-935-8666

Anne's Dolls, Marvin's Toys
13629 Victory Blvd.
Van Nuys, CA 91401
818-785-1177

Area 51 Toys & Collectables
508 Wayne Dr.
Wilmington, NC 28403
910-762-2553

Atomic Age
318 E. Virginia Rd.
Fullerton, CA 92631
714-446-0736

Banowetz Antique Mall
122 McKinsey Ave.
Maquoketa, IA 52060
319-652-2359

Blast from the Past
210 S. First St.
Ann Arbor, MI 48104
313-913-8890

Blue Comet Hobby
12 Meeker Ave.
South Amboy, NJ 08879
908-583-8637

Bob Sellstedt
9307 Hillingdon Rd.
Woodbury, MN 55125
612-738-1597

Bojo
P.O. Box 1403
Cranberry Twp, PA 16066
412-776-0621

Boom Gallery
146 N. Northwest Hwy.
Park Ridge, IL 60068
847-823-2666

Brian Razzi
36 Ingate Terrace
Baltimore, MD 21227
410-536-7147

Cartoon World
11054 Ventura Blvd.
Studio City, CA 91604
213-878-0611

Casey's Collectible Corner
HCR Box 31, Rte. 30
N. Blenheim, NY 12131
607-588-6464

Childhood Memories
P.O. Box 5367
Youngstown, OH 44514
216-757-8279

Christiane Swain
74 Ranney Corner Rd.
Ashfield, MA 01330
413-628-3213

Cincinnati Sci-Fi
2312 Kemper Ln. #2
Cincinnati, OH 45206
513-729-9533

Collectible Toys
P.O. Box 329
Sun City, CA 92586
714-672-9502

Collectibles Madness
483 Federal Rd.
Brookfield, CT 06804
203-740-9472

Collection Connection
P.O. Box 18552
Hamilton, OH 45018
513-851-9217

Collectorholics
15006 Fuller Ave.
Grandview, MO 64030
816-322-0906

Continental Hobby House
P.O. Box 193
Sheboygan, WI 53082
414-693-3371

Cortrade International
419 Main St., Suite 246
Huntington Beach, CA 92648
800-290-9069

Cosmic Cargo
1296 E. Gibson Rd. #264
Woodland, CA 95776
916-661-2384

Cricket Hill Toys
1175 Co. Rt. 1
Oswego, NY 13126
315-343-6715

Damn Yankees
P.O. Box 126
Saint Augustine, FL 32085
904-824-3336

David Hay
6363 Birch Leaf Ct.
Burke, VA 22015
202-287-2136

David Welch
P.O. Box 714
Murphysboro, IL 62966
618-687-2282

DDT
3970 S. Leavitt Rd. SW
Warren, OH 44481
216-394-0442

Dennis Shoup
1119 S. Curtis Ave.
Kankakee, IL 60901
815-932-0230

Dewayne or Lora Bryner
P.O. Box 1911
Denver, DE 19903
302-734-7018

Dr. Tongue's 3-D House of Toys
1408 E. Burnside St.
Portland, OR 97214
503-235-9849

Ed and Ann Risner
700 S. Union St.
Fostoria, OH 44830
419-435-2124

Ed Janey
1756 65th St.
Garrison, IA 52229
319-477-8888

Emerald City Comics and Collectibles
9249 Seminole Blvd.
Seminole, FL 34642
813-398-2665

Everything Under the Sun
P.O. Box 5607
Evanston, IL 60204
312-381-9941

Frantic City Toys
4232 Danor Dr.
Reading, PA 19605
610-929-8859

Fred Lambert
P.O. Box 7713
East Rutherford, NJ 07073
201-438-6047

Frederick Ross
2128 North 38th St.
Milwaukee, WI 53208
414-444-5836

G.A. Corp.
P.O. Box 572
Nutley, NJ 07110
201-935-3388

Gasoline Alley
6501 20th Ave. NE
Seattle, WA 98115
206-524-1606

Grandstand Collectibles
6357 Lampkins Bridge Rd.
College Grove, TN 37046
615-395-4108

Greg Roccaro
894 Armstrong Ave. Apt. 2-4
Staten Island, NY 10308
718-967-8345

Gregory Hamm
8 Westwind Ct.
Hawthorn Wood, IL 60047
708-540-8318

Hake's Americana and Collectibles
P.O. Box 1444
York, PA 17405
717-848-1333

Halloween Outlet
246 Park Ave.
Worcester, MA 01609
800-425-9336

Harry Klaes
2165 W. Avenue 134th
San Leandro, CA 94577
510-614-0744

Hobbies-On-Line
10120 Main St, Apt. B
Clarence, NY 14031
716-759-7151

Hollywood Legends
6621A Hollywood Blvd.
Los Angeles, CA 90028
213-962-6742

Hot Comics & Collectables
7971 Southtown Ctr.
Bloomington, MN 55431
612-881-4799

House of Cool Toys
P.O. Box 1014
Canyon Country, CA 91386
805-298-8697

Howdy-Do
72 E. 7th St.
New York, NY 10003
212-979-1618

I Need Moore Toys
1243 Bay Area Blvd.
Houston, TX 77058
713-480-4928

International Executive
954-572-3493

Jamar Classic Toys
11939 Manchester Rd.
Des Peres, MO 63131
314-822-7202

Jamar Company
5015 State Rd.
Medina, OH 44256
216-239-2889

James Koval
P.O. Box 1700
Plainfield, NJ 07061
908-757-0828

Jeff Lowe
5005 Tamara Ln.
West Des Moines, IA 50265
515-226-9404

Jerry Harnish
110 Main St.
Bellville, OH 44813
419-886-4782

John DiCicco
57 Bay View Dr.
Shrewsbury, MA 01545
508-797-0023

Jordan Corp.
9797 W. Kolfax 1D
Lakewood, CO 80215
303-239-8229

Just Kids Nostalgia
310 New York Ave.
Huntington, NY 11743
516-423-8449

Klingenberg's
4217 Fernbrook Dr.
Loveland, CO 80537
303-669-4638

Kool Toys & Collectibles
4624 Steepleton Way
Charlotte, NC 28215
704-394-1066

Legends & Heroes
14968 E. Wagontrail Pl.
Aurora, CO 80015
303-617-0052

Lin Terry
1000 Airport Rd., Suite 103
Lakewood, NJ 08701
908-370-5252

Louis Turczynski
657 Furnace Rd.
Wernersville, PA 19565
610-670-1965

Lyle Rhodebeck
7920 Stemen Rd.
Pickerington, OH 43147
614-837-7570

M & J Variety
802 1st Ave.
Elizabeth, NJ 07201
908-820-0082

Mad About Toys
1525 Aviation Blvd.
Redondo Beach, CA 90278
310-318-1829

Manochio Enterprises
P.O. Box 2010
Saratoga, CA 95070
408-996-1963

Mark Clark
P.O. Box 2515
Liverpool, NY 13089
315-457-3139

Marz Toys & Collectibles
2658 E. Fowler Ave.
Tampa, FL 33612
813-971-8686

Melanie Kiviat
P.O. Box 24531
Los Angeles, CA 90024
310-274-2874

Metropolis Collectibles
873 Broadway, Suite 201
New York, NY 10003
212-260-4147

Michael Melito
29 Foodmart Rd.
Boston, MA 02118
617-662-2189

Midwest Memorabilia
P.O. Box 261126
Highlands Ranch, CO 80163
303-683-8883

Mike's Melodies & Memories
2239 A. Bloor St. West
Toronto, Ontario, Canada M6S IN7
416-763-7298

Nostalgic Americana
3067 E. El Monte Way
Fresno, CA 93721

Nostalgic Toy Creations Ltd.
P.O. Box 441
Anamosa, IA 52205
319-462-4626

Nostalgic Toys
P.O. Box 470887
Los Angeles, CA 90047
213-755-0311

Peddler, The
2104 W. Springfield Ave.
Champaign, IL 61821
217-351-9447

Planet Toys
429 Bruce Ave.
Stratford, CT 06497

Randy Aguiar
12 E. Public St.
Assonet, MA 02702
508-644-3362

Relix International
P.O. Box 98
Brooklyn, NY 11229
718-258-0009

Remembrance of Innocence
P.O. Box 628
Damariscotta, ME 04543
207-563-2421

Robert Swerdlow
220-55 46th Ave.
Bayside, NY 11361
718-224-2599

Rocky Mountain Trading Card
P.O. Box 460490
Aurora, CO 80046
303-680-6233

Ron Hill
905 El Oro Dr.
Auburn, CA 95603
916-888-0291

Scooter's Custom Works
843 Rocklyn Dr.
Pittsburgh, PA 15205

Searcher, The
34064 White Oak Ln.
Gurnee, IL 60031
708-855-8630

Shadowhouse Collectibles
P.O. Box 2284
Wilsonville, OR 97070
503-691-5785

Spastic over Plastic
288 Parker Ave.
Clifton, NJ 07011
201-772-5466

Stacey's TV Toys
5930 S. 32nd St.
Lincoln, NE 68516
402-423-1904

Steve Kaplan
P.O. Box 2391
New York, NY 10163
212-687-2419

Steve's Lost Land of Toys
3572 Turner Ct.
Fremont, CA 94536
510-795-0598

Steve's Quality Collectibles
435 E. 7th St.
Dover, OH 44622
216-343-8704

Steven Agin
P.O. Box 687
Delaware, NJ 07833
908-475-1796

Super Collector
16547 Brookhurst St.
Fountain Valley, CA 92708
714-839-3693

Tally Ho Studio
639 Park Ave. SW
Canton, OH 44706
216-452-4488

Terri's Toys
419 S. 1st St.
Ponca City, OK 74601
405-762-5174

That Toy Guy
502 East Baltimore Pike
Lansdowne, PA 19050
610-394-8697

That's Entertainment
P.O. Box AA
Reamstown, PA 17567
717-335-3100

Thor Uranis
26512 Wick Rd.
Taylor, MI 48180
313-292-3567

Tin Sedan
12 N. Main St.
Smyrna, DE 19977
302-653-3535

Tom Knowlton
P.O. Box 6283
Rochester, MN 55903
507-288-8441

Tom Wright
28 Wade Ave.
Baltimore, MD 21228
410-828-5481

Toonerville Antiques
6118 Phinney Ave. N.
Seattle, WA 98103
206-706-7037

Toy Loft
P.O. Box 723
Florence, KY 41022
606-525-7032

Toy Pirates
99042 46th Pl. W.
Mukilteo, WA 98275
206-348-5588

Toy Seductions
19 Rock Hill Rd. #3B
Bala Cynwyd, PA 19004
610-660-8204

Toys Around the Clock
3214 213th St.
Flushing, NY 11361
718-229-0041

Trader Jacks
61 S. State Rd. 7 & 441
Fort Lauderdale, FL 33317
954-583-3334

Two Guys Toys
P.O. Box 598
Feasterville, PA 19053
215-364-6685

Uniquely Different
104 Park Ave.
Mechanicville, NY 12118
518-664-6966

Village Idiot
21 Wyoming Ave.
Billings, MT 59101
406-248-4875

Vintage Toy Depot
P.O. Box 206
Kenilworth, NJ 07033
908-276-5464

Wayback Machine
P.O. Box 215
Hope, RI 02831
401-828-7775

Wesley's Inheritance
R.D. #1, Box 140
Hallstead, PA 18822
717-967-2846

Whiz Bang Collectibles
P.O. Box 300546
Fern Park, FL 32730
407-260-8869

Wild Pig Antique Toys
P.O. Box 187
East Winthrop, ME 04343
207-395-4419

World Collectibles Center
18 Vesey St.
New York, NY 10007
212-267-7100

X-Head Un-Ltds
383 Route 306
Monsey, NY 10952
914-362-3605

Yankee Peddler Antiques and Collectibles
5205B Greenock Rd.
Lothian, MD 20711
410-741-9814

Yestertoys, Inc.
240 Pawtuxet Ave.
Cranston, RI 02905
401-467-6233

Your Old Friends Doll Shop
21 5th St. SW
Mason City, IA 50401
515-424-0984

Zack Isaacs
4408 16th Ave.
Brooklyn, NY 11204
718-438-0700

Zippy's
P.O. Box 800174
Roswell, GA 30075
770-998-8641

Character Toys, Wanted

Car 54, Where Are You?
P.O. Box 110413
Nashville, TN 37222

Mark Kalman
P.O. Box 44098
Van Nuys, CA 91412
818-997-0250

The Beatles Mobile Museum
507 Normal Ave.
Normal, IL 61761
309-452-9376

Comics, For Sale

Camelot Comics
1102 Beulah Ave.
Tylertown, MS 39667
601-731-8004

Comic Stop
12 Riverbend Dr.
Rome, GA 30161
706-295-1444

Comics, Wanted

Duane Wetzel
317 N. 3rd St.
Coopersburg, PA 18036
610-282-4082

Construction & Earth Moving Machinery

Buffalo Road Hobby Imports / Die-Cast
10120 Main St.
Clarence, NY 14031
716-759-7451

Ralph E. Garber Enterprises
1351 Copenhaffer Rd.
Dover, PA 17315
717-266-5348

Ten Oaks Toys
P.O. Box 133
Glenelg, MD 21737
410-489-5341

Corgi Toys

ACME Toy Co.
9 Station Rd., Erdington
Birmingham, B236UB ENGLAND

Cracker Jack & Radio Prizes

Ron Toth
72 Charles St.
Rochester, NH 03867
603-335-2062

Disneyana

Jack Kehoe Distributing
800 S. Larchwood Dr.
Brea, CA 92821
714-990-0918

Doll Furniture

Bobbi Segal
P.O. Box 39
Julian, PA 16844
814-355-2542

Dolls For Sale, Barbie

Barbara Weaver
14924-258 Ave. SE
Issaquah, WA 98027
206-391-1332

Doll & Hobby Shoppe
203 N. Woodland Blvd.
Deland, FL 32720
904-734-3202

Heavenly Doll
P.O. Box 774
South Lake Tahoe, CA 96156
702-588-3655

Hello Dollies
2564 Wigwam Parkway, Ste 206
Henderson, NV 89014
702-897-7624

Marl and B
10301 Braden Run
Bradenton, FL 34202
941-751-6275

Michele Prusinowski Dolls
106 E. DeKalb Pike
King of Prussia, PA 19406
610-279-2055

Dolls For Sale, Others

Desiree's Dolls and Toys
16-1/2 W. College Ave.
Westerville, OH 43081
614-890-6424

Lin Murphy
1089 Granby Rd.
Chicopee, MA 01020
413-533-1289

Dolls Wanted

Ann Cheney
16 Sagamore Ave.
Winthrop, MA 02152
617-539-1057

The Cottage Couturiere for Fashion Dolls
68 Washington Rd.
Sayreville, NJ 08872
908-613-1836

Ertl Toys

Toys for the Boys
213 Bridge St.
LeSueur, MN 56058
800-847-6312

Farm Vehicles, Implements

Dick's Toys
Rt 3 Box 318
Napleton, MN 56065
507-524-3275

Fast Food Collectibles

Bill & Pat Poe
220 Dominica Circle E
Niceville, FL 32578
904-897-4163

Figurines & Statues

David R. Calderone
137 Ridgewood Dr.
Freedom, PA 15042
412-869-0642

Fire Engines Wanted

David's Fire Toys
8861 E. Jenan Dr.
Scottsdale, AZ 85260
602-860-1486

G.I. Joe

Comic Collection, Toys and Compact Discs
931 Bustleton at Street Rd.
Feasterville, PA 19053
215-357-3332

Cotswold Collectibles
P.O. Box 249
Clinton, WA 98236
216-744-1552

Elvis Rodriguez
16 Harrison Ave.
Coram, NY 11727
516-736-8730

James DeSimone
150 S. Glenoaks Blvd.
Burbank, CA 91510
818-563-1179

Kevin B. Bolger
510 N. Prospect Manor Ave.
Mount Prospect, IL 60056

Tripwire Toys
702 Mangrove Ave., Suite 125
Chico, Ca 95926
916-877-7846

Games, Board & Boxed

Lyle Rhodebeck
431 Maranatha Dr.
Marysville, OH 43040
513-642-1318

Hot Wheels Toys

Bob Goforth
4061 E. Castro Valley Blvd.
Castro Valley, CA 94552
510-889-6676

Brownie Brown
3680 W. Sylvania St.
Springfield, MO 65807
417-887-8469

Cars and Coins
4942 US Hwy 19
New Port Richey, FL 34652
813-848-3626

Joseph Verkler
312 S. 6th
Dekalb, IL 60115
815-756-7412

Mr. Lloyd
P.O. Box 5140
Springfield, IL 62705

Lunch Boxes

Jerry Magee
2121 Leech Ave.
Sioux City, IA 51107
712-277-8916

Rob Bender
1809 7th Ave. #1008
Seattle, WA 98101
206-522-9896

Magic Tricks & Sets

Phil Temple, Magic Set Collectors
P.O. Box 561
Novato, CA 94948
415-897-5130

Matchbox Toys

Sonny's Toys & Collectibles
Hendel Bldg. Route 27
Old Mystic, CT 06372
860-536-0646

Biff Smith
1344 W. Flower
Phoeniz, AZ 85013
602-274-9366

Joseph Saine
P.O. Box 50506
Toledo, OH 43605
419-691-0008

Only Mint
289 Mount Hope Ave.
Dover, NJ 07801
201-366-4677

Model Kits

Art 'N Things
133 S. Anderson
Fairview, NJ 07022
201-943-2288

Castle Comics & Collectibles
396 Bridgeport Ave.
Milford, CT 06460
203-877-3610

Character Shop, The
684 Broadway
Everett, MA 02149
617-387-0660

Club Daikaiju
P.O. Box 1614
Fort Lee, NJ 07024
201-587-8112

Creative Three Hobbies
7475 Miller Dr.
Bath, PA 18014
610-837-8582

Harry's Hobbies & Supply
12922 Harbor Blvd. #832
Garden Grove, CA 92840
714-563-6606

James Crane
15 Clemson Ct.
Newark, DE 19711
302-738-6031

Monsters in Motion
330 E. Orangethorpe Ave.
Placentia, CA 92870
714-577-8863

Motorcycles

Moto-Mini
RR 4, Box 516A
McKinney, TX 75050
214-837-2686

Movie & TV Show Toys

D.C. Hollis
P.O. Box 65
Mount Tabor, NJ 07878

Hollywood Dream Factory
1842 W. Sylvania Ave.
Toledo, OH 43613
419-474-3065

Jim's TV Guides and Collectibles
P.O. Box 4767
San Diego, CA 92164
619-462-1953

M.S. Oliver
36 MacArthur Blvd.
Danvers, MA 01923
508-774-3955

PH Enterprises
1923 3rd Concession
Windsor, Ontario, Canada N9E
519-969-3056

Queen's Collection
Carren Kopleton
P.O. Box 235
Clifton, NJ 07011
201-772-8548

Rialto Movie Art
81-1/2 S. Washington St.
Seattle, WA 98104
206-622-5130

Stephen Russo
P.O. Box 150, Revere Branch
Boston, MA 02151

Video Resources N.Y.
220 West 71st St.
New York, NY 10023
212-724-7055

Non-Toy Misc.

Old Country Barn
1554 Holly Pike
Carlisle, PA 17013
717-245-9303

Second Stop Shop
6215 Chambersburg Rd.
Fayetteville, PA 17222
717-352-4762

Pedal Cars

Juvenile Automobiles
P.O. Box 221
Sheldonville, MA 02070
401-766-9661

Pedal Car Graphics
1207 Charter Oaks Dr.
Taylors, SC 29687
864-244-4308

Playsets

Specialty Publishing
P.O. Box 1355
La Crosse, WI 54601
608-781-1894

William M. Staley
316 Edgewood Ridge Rd.
Wilkesboro, NC 28697
910-667-7185

Publications

Action Figure News
556 Monroe Tpke.
Monroe, CT 06468
203-452-7286

The Premium Watch Watch
Sharon Iranpour
24 San Rafael Dr.
Rochester, NY 14618
716-381-9467

Tomart Publications
3300 Encrete Ln.
Dayton, OH 45439
513-294-2250

Toy Trucker & Contractor
7496 106 Ave. S.E. (Dept. TSD)
LaMoure, ND 58458-9404
800-533-8293

Railroads, Electric

The Cola Freightyard
Downtown Plaza, 100 Brown St #14
Middleton, PA 17057
717-944-2336

Trains & Toys
4352 S. 42 St.
Omaha, NE 68107
402-734-1426

Railroads, Others

Merri-Seven Trains
2097 Tucker Dr.
Troy, MI 48098
810-474-5373

Robots

Mark Bergin
2 E. Mountain Rd.
Peterborough, NH 03458
603-924-2079

Robert Johnson
2720 E. 50th St.
Minneapolis, MN 55417
612-721-5256

Stanley Luksenburg
21475 Lorain Rd.
Fairview Park, OH 44126
216-356-1118

Tin Lilliput Toys
P.O. Box 711
Kirkland, WA 98083
306-823-9530

Space Toys

Brandon Helwig
1240 Golfside Drive
Winter Park, FL 32792
407-673-3726

Chris Smiths Works
P.O. Box 27071
Oakland, CA 94602
510-530-1673

Collectibles Craze
39 Dorchester Dr.
East Brunswick, NJ 08816
908-613-0907

Federation Supply Base
303 Concord Ct.
Huron, OH 44839
419-433-6670

Galactic Trading Post
1660 S. Albion St., Suite 1004
Denver, CO 80222
303-756-3377

Lightside Space Collectibles
20201 Burnt Tree Lane
Walnut, CA 91789

Midwest Memorabilia Inc.
P.O. Box 261126
Highlands Ranch, CO 80163
303-805-1210

MSC Toys & Comics
16-1/2 W. College Ave.
Westerville, OH 43081
614-890-6547

Spiral
Kondokata 1-30-7
Tokyo, Japan
003-470-0603

William Renshaw
7621 U. Dr. S, Box 328
Union City, MI 49094
517-741-7805

Sports Related, For Sale

Penny Lenny
37 Hudson St.
Yonkers, NY 10701

Toy Parts, For Sale

Thomas Toys
P.O. Box 405
Fenton, MI 48430
810-629-8707

Toy Repairs

Battery-Operated Toy Repair
2825 S. County Rd. 275W
New Castle, IN 47362
317-987-7507

C. Wis. Greatest Toys on Earth
Hobbie City, 1238 S. Beach Blvd.
Anaheim, CA 92804
714-995-4151

Toy Doctor, The
RR 1 Box 202 Eades Rd.
Red Creek, NY 13143
315-754-8846

Toys For Sale, Misc.

Attic Busters
415 Dodge Ave.
Jefferson, LA 70121
504-733-0676

Blue Moon Collectibles
3832 14th Ave.
Rock Island, IL 61201
309-786-6049

Collectiques
820 Mocksville Ave.
Salisbury, NC 28144
704-638-9406

Curious Goods
71 W. Main St.
Westminster, MD 21157
410-840-8750

Cynthia L. Case
11324 Knollridge Ct. #3
Indianapolis, IN 46229

Days Gone By
3200 NE 43rd Place
Ocala, FL 34479
352-622-8908

Eggland Toys
408 W. 36th St., Apt. 3-D
New York, NY 10018
212-643-9760

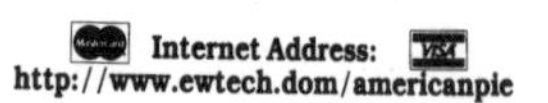

Flying Saucers
P.O. Box 1432
Bisbee, AZ 85603
520-432-4858

Hidden Treasures
234 W. Main St.
Loudenville, OH 44842
419-938-6587

Hog Wild
100 E. North Loop #A
Austin, TX 78751
512-467-9453

Housatonic Comics & Collectibles
P.O. Box 703-B, Rt. 7
Gaylordsville, CT 06755
860-350-1921

Hurricane Town Collectibles
1232 Stevens Lane
Mobile, AL 36618
334-602-8164

J & J Antiques and Collectables
160 Central Ave.
West Keansburg, NJ 07734
908-495-9045

Janet Giliberty
1989 60th N.
Saint Petersburg, FL 33710
813-343-2264

Kittles Hobby Shop
188 2nd St.
Barberton, OH 44203
330-753-9555

Mad Stasher's Delectable
Collectables
114 S. Main St. Route 19
Belfast, NY 14711
716-365-8191

Patrick Hausenfleck
7714 Camellia Ln.
Stockton, CA 95207
209-474-6483

Paul Burkett
619 Yoder St.
Johnstown, PA 15901
814-535-4845

Purple Dragon Gifts
1977 N. Olden Ave.
Trenton, NJ 08618
609-883-6683

Re-Play Toys & Collectibles
2153 W. Busch Blvd.
Tampa, FL 33612
813-931-5677

Toys Wanted, Misc.

Charles Anderson Co.
30 N. First St.
Minneapolis, MN 55401
612-339-5181

Justin's Toys
329 S. Mayfair Ave. #305
Daly City, CA 94015
415-994-1514

Trading Cards

Barrington Square
2322 W. Higgins Rd.
Hoffman Estates, IL 60195
708-882-7080

H & L Card Shop
534 Gates St.
Aurora, IL 60505
708-851-0177

Jim Walls
3310 Wiltshire Dr.
Avondale Estates, GA 30002
404-296-2480

Ko's Cards & Collectibles
585 Remsen Ave.
Brooklyn, NY 11236
718-381-1868

Trucks for Sale

Bruce Hemmah
100 Main St. W.
Cannon Falls, MN 55009
507-263-4067

G & J Collectibles
118 Candora Rd.
Maryville, TN 37804
423-983-7187

Homestead Collectibles
P.O. Box 173
Mill Hall, PA 17751
717-726-3597

JMT Replicas
P.O. Box 1105
Bartlesville, OK 74005
918-336-7481

Limited Edition Toys
60 Whitney Dr.
Valatie, NY 12184
518-758-6103

Lloyd J. Wenger
831 Hilltop Rd
Myerstown, PA 17067
717-866-7147

Midnight Express
P.O. Box 1854
Melrose Park, IL 60101
708-354-0950

Newport Precision Trucks
1560 Placentia Ave. #B-10
Newport Beach,CA 92663
714-646-2502

Smith-Miller Inc.
P.O. Box 139
Canoga Park, CA 91305
818-703-8588

Transport Collectables
607 Kulp Rd. RR4
Pottstown, PA 19465
610-323-3114

Vehicles, Die-Cast and Plastic

Die-Cast Direct
1009 Twilight Trail
Frankfort, KY 40601
502-227-8697

Mars Collectibles
155 Big Elk Mall
Elkton, MD 21921
410-398-3636

Toy Time
121 Laurel St.
Fitchburg, MA 01420
508-827-5261

Vehicles, Tin & Steel

Trader Fred's Toys of Yore
Rt. 132, Box 65
Thetford Center, VT 05075
802-785-2845

Wind-up & Battery Toys

Lee Mitchell
175 E. Delaware
Chicago, IL 60611
312-337-3123

Peter Adduci
64 H St.
Hull, MA 02045

Steve Fisch
10 W. Main St.
Wappingers Fl., NY 12590
914-298-8882

1997 Toy Show Calendar

ALABAMA

Jan 11 AL, Birmingham. Antique & Collectible Toy & Doll Show. Birmingham-Jefferson Civic Ctr. South Hall on 9th Ave. N. Southeastern Collectible Expo, 1819 Peeler Rd., Vale, NC 28168. PH: 704-276-1670.
Mar 8 AL, Fairhope. "March Fantasy Doll Show". Civic Center Auditorium. SH: 9:30am-4pm. Doris Perdue, PO Box 386, Montrose, AL 36559. PH: 334-928-0855.

ARIZONA

Feb 23 AZ, Phoenix. "Barbie Madness Mega Show". Red Lion Hotel, Laposada, 949 E. Lincoln Dr. SH: 10am-4pm. PH: 602-952-0420.
Mar 16 AZ, Tucson. Collectible Toy Show. Marketplace USA, 3750 E. Irvington Rd. SH: 9am-3pm, T: 300. Lou Pariseau, 1111 E. Limberlost Rd. #164, Tucson, AZ 85719. PH: 520-293-3178.

ARKANSAS

May 17 AR, Eureka Springs. 15th Annual Doll & Toy Show. Convention Ctr., behind Inn of the Ozarks, Rt. 62 W. SH: 10am-4pm, T: 90. Marilyn Bromstad, Rt. 7, Box 574C, Eureka Springs, AR 72631. PH: 501-253-2244.
Sep 13 AR, Eureka Springs. 10th Annual Toys, Dolls & Trains Show. Convention Ctr., behind Inn of the Ozarks Best Western GZW. SH: 9am-4pm, T: 75. George Bromstad, Rt. 7, Box 574C, Eureka Springs, AR 72631. PH: 501-253-9989 or 253-2244 eves.

CALIFORNIA

Jan 4-5 CA, Pasadena. Toyrific Antique & Collectible Toy Show. Pasadena Ctr., 300 E. Green St. SH: 10am-5pm. Mike Stannard, PO Box 2037, San Bernardino, CA 92406. PH: 909-880-8558.
Jan 4-5 CA, Santa Clara. NNL Western Nationals West Coast Model Expo. Convention Ctr., 5001 Great America Pky. SH: Sat. 10am-8pm, Sun. 10am-3pm, T: 125. Rex Barden, PO Box 149, Santa Clara, CA 95050. PH: 408-371-3762.
Jan 5 CA, San Mateo. Toy Show. County Expo Ctr., 2495 S. Delaware St. SH: 10am-3pm, T: 200. Dugger-Hill Promos., PO Box 243, Auburn, CA 95604. PH: 916-888-0291 or 408-476-8612.
Jan 5 CA, San Bernardino. Outdoor Market. National Orange Showgrounds. PH: 213-560-SHOW ext. 12 or 909-795-1884.
Jan 11-12 CA, San Diego. Linda's Teddy Bear, Doll & Antique Toy Show. Scottish Rite Center, 1895 Camino Del Rio South. SH: Sat. 10:30am-4pm, Sun. 10am-3pm, 150 dealers. Linda Mullins, PO Box 2327, Carlsbad, CA 92018. PH: 619-434-7444. FAX: 619-434-0154.
Jan 11 CA, Santa Rosa. Winter Doll Show. Sonoma Cty. Fairgrounds, Hwy. 101 to Hwy. 12 E. to first offramp (South E St.). SH: 10am-4pm. Golden Gate Shows, Fern Loiacono, PO Box 1208, Ross, CA 94957. PH: 415-459-1998 or FAX: 415-459-0827.
Jan 12 CA, Pasadena. Flea Market. Rose Bowl. PH: 213-560-SHOW ext. 11.
Jan 12 CA, Pasadena. Arts & Crafts Outdoor Festival. Rose Bowl. PH: 213-560-SHOW ext. 44.
Jan 12 CA, San Bernardino. Outdoor Market. National Orange Showgrounds. PH: 213-560-SHOW ext. 12 or 909-795-1884.
Jan 16-19 CA, San Jose. Pinnacle NHL Fantasy All-Star Weekend. McEnery Conv. Ctr. SH: Thurs. 6pm-10pm, Fri. 9am-10pm, Sat. & Sun. 9am-6pm. NHL Enterprises, L.P., Schulte Marketing, Inc., 217 E. 85th St., Ste. 266, New York, NY 10028. PH: 212-772-8605 or FAX: 212-879-8358.
Jan 18 CA, San Mateo. Doll & Teddy Bear Show. Cty. Expo Ctr. SH: 10am-3:30pm. Lloyd Hogan, 5621 Glencrest Ln., Orangevale, CA 95662. PH: 916-989-9291.
Jan 18-19 CA, Glendale. All-American Collectors Show. Civic Auditorium, 1401 N. Verdugo Rd. SH: Sat. 12noon-7pm, Sun. 10am-4pm. PH: 310-455-2894 or 818-980-5025.
Jan 19 CA, San Jose. "Barbie Madness Mega Show". Red Lion Hotel, 2050 Gateway Pl. SH: 10am-4pm. PH: 408-453-4000.
Jan 19 CA, San Bernardino. Outdoor Market. National Orange Showgrounds. PH: 213-560-SHOW ext. 12 or 909-795-1884.
Jan 19 CA, Stanton. NASCAR, NHRA, Indy, Hot Wheels & Johnny Lightning Coll. Show. Rack & Cue Billiards, 7520 Katella Ave. near Western. SH: 9am-2:30pm. Ron, PH: 909-902-0217.
Jan 24-26 CA, San Diego. Prestigious Antiques & Collectibles Show. Scottish Rite Center, 1895 Camino Del Rio South. SH: Fri. 12noon-9pm, Sat. 12noon-7pm, Sun. 12noon-5pm. Sy Miller Productions, PO Box 967, Rancho Santa Fe, CA 92067. PH: 619-756-3275.
Jan 24 CA, National City. San Diego Toy Show. 1545 Tidelands Ave., Suite "J", I-5 S., take R. on 24th St. Exit. SH: 6pm-9pm, T: 25. Edward Morris, PH: 619-441-8198.
Jan 25 CA, Mission Viejo. Toy Collectibles Show. Capistrano Valley H.S., 26301 Via Escolar. SH: 9am-2pm, T: 80. Stephen Spero, PH: 714-768-8901.
Jan 26 CA, San Bernardino. Outdoor Market. National Orange Showgrounds. PH: 213-560-SHOW ext. 12 or 909-795-1884.
Feb 1 CA, San Jose. Winter Doll Show. Santa Clara Cty. Fairgrounds, Hwy. 101 to Tully Rd. Exit West. SH: 10am-4pm. Golden Gate Shows, PO Box 1208, Ross, CA 94957. PH: 415-459-1998 or FAX: 415-459-0827.
Feb 2 CA, Baldwin Park. 14th Annual Barbie Show. Marriott, 14635 Baldwin Park Town Ctr. SH: 10am-3pm. Inge Strong, PH: 818-331-4695 or Sharon Mairs, PH: 818-444-3734.
Feb 2 CA, Ventura. Flea Market & Swap Meet. Cty. Fairgrounds, Seaside Park. PH: 213-560-SHOW ext. 13.
Feb 2 CA, San Bernardino. Outdoor Market. National Orange Showgrounds. PH: 213-560-SHOW ext. 12 or 909-795-1884.
Feb 8 CA, Pleasanton. Doll Show & Winter Bear Jamboree. Alameda Cty. Fairgrounds. SH: 9am-4pm. Lloyd Hogan, 5621 Glencrest Ln., Orangevale, CA 95662. PH: 916-989-9291.
Feb 9 CA, Pasadena. Flea Market. Rose Bowl. PH: 213-560-SHOW ext. 11.
Feb 9 CA, Pasadena. Arts & Crafts Outdoor Festival. Rose Bowl. PH: 213-560-SHOW ext. 44.
Feb 9 CA, San Bernardino. Outdoor Market. National Orange Showgrounds. PH: 213-560-SHOW ext. 12 or 909-795-1884.
Feb 15-16 CA, Ventura. Dolls, Bears, Supplies & Crafts Show. Fairgrounds-Commercial Bldg., 10 W. Harbor Blvd. #101, Calif. St. or Ventura Ave. Exits. SH: Sat. 10am-5pm, Sun. 11am-4pm. The Miller Prod. Group, PO Box 967, Rancho Santa Fe, CA 92067. PH: 619-756-3275.
Feb 15 CA, San Diego. 21st Annual Doll Show. Scottish Rite, 1895 Camino del Rio S., Exit I-8 at Mission Center Rd. SH: 10am-4pm. Richard Weed, PH: 619-945-9397, Bettie Vasey, PH: 619-283-9283 or Peg Fulwider, PH: 619-444-6509 dealers only.
Feb 16 CA, Santa Ana. Toy Gun Collectors Round-Up. Elks Club, 212 Elk Ln. SH: 9am-3pm, T: 80. Toy Gun Round-Up, Jim Buskirk, 3009 Oleander Ave., San Marcos, CA 92069. PH: 619-599-1054 or Chuck Quinn, PH: 800-585-2392.
Feb 16 CA, San Bernardino. Outdoor Market. National Orange Showgrounds. PH: 213-560-SHOW ext. 12 or 909-795-1884.
Feb 22 CA, Mission Viejo. Toy Collectibles Show. Capistrano Valley H.S., 26301 Via Escolar. SH: 9am-2pm, T: 80. Stephen Spero, PH: 714-768-8901.
Feb 23 CA, San Bernardino. Outdoor Market. National Orange Showgrounds. PH: 213-560-SHOW ext. 12 or 909-795-1884.
Feb 28 CA, National City. San Diego Toy Show. 1545 Tidelands Ave., Suite "J", I-5 S., take R. on 24th St. Exit. SH: 6pm-9pm, T: 25. Edward Morris, PH: 619-441-8198.
Mar 1-2 CA, Victorville. Dolls, Bears, Supplies & Crafts Show. Fairgrounds, 14800 7th St. SH: Sat. 10am-5pm, Sun. 11am-4pm. Miller Production Group, PO Box 967, Rancho Santa Fe, CA 92067. PH: 619-756-3275.
Mar 2 CA, Ventura. Flea Market & Swap Meet. Cty. Fairgrounds, Seaside Park. PH: 213-560-SHOW ext. 13.
Mar 2 CA, San Bernardino. Outdoor Market. National Orange Showgrounds. PH: 213-560-SHOW ext. 12 or 909-795-1884.
Mar 2 CA, Sacramento. Antique Toy Show. Scottish Rite Center, 6151 H St. SH: 10am-3pm, T: 100. Wayne Gateley, PO Box 221052, Sacramento, CA 95822. PH: 916-448-2655.
Mar 8 CA, San Rafael. Spring Doll Show. Marin Cty. Civic Center Exhibition Hall, Hwy. 101 to N. San Pedro Rd., Exit East. SH: 10am-4pm. Golden Gate Shows, Fern Loiacono, PO Box 1208, Ross, CA 94957. PH: 415-459-1998 or FAX: 415-459-0827.
Mar 9 CA, Costa Mesa. 8th Annual West Coaster Toy Soldier & Miniature Figure Show. Red Lion Hotel (Orange County Airport), 3050 Bristol St. SH: 9am-3pm, T: 200. PH: 619-758-5481.
Mar 9 CA, Pasadena. Flea Market. Rose Bowl. PH: 213-560-SHOW ext. 11.
Mar 9 CA, Pasadena. Arts & Crafts Outdoor Festival. Rose Bowl. PH: 213-560-SHOW ext. 44.
Mar 9 CA, San Bernardino. Outdoor Market. National Orange Showgrounds. PH: 213-560-SHOW ext. 12 or 909-795-1884.
Mar 16 CA, Pomona. Toyrific Antique & Collectible Toy Show. Fairplex, L.A. Cty. Fair, Hotel & Exposition Complex. SH: 10am-4pm. 20th Century Prods., PO Box 2037, San Bernardino, CA 92406. PH: 909-880-8558 or FAX: 909-880-8096.
Mar 16 CA, San Mateo. All-Barbie Show. County Expo Ctr. SH: 10am-3:30pm. Lloyd Hogan, 5621 Glencrest Ln., Orangevale, CA 95662. PH: 916-989-9291.
Mar 16 CA, San Mateo. Toy Show. County Expo Ctr., 2495 S. Delaware St. SH: 10am-3pm, T: 200. Dugger-Hill Promos., PO Box 243, Auburn, CA 95604. PH: 916-888-0291 or 408-476-8612.
Mar 16 CA, San Bernardino. Outdoor Market. National Orange Showgrounds. PH: 213-560-SHOW ext. 12 or 909-795-1884.
Mar 22-23 CA, Costa Mesa. Dolls, Bears, Supplies & Crafts Show. Orange Co. Fair & Exposition Center, 88 Fair Dr. SH: Sat. 10am-5pm, Sun. 11am-4pm. Miller Production Group, PO Box 967, Rancho Santa Fe, CA 92067. PH: 619-756-3275.
Mar 23 CA, San Bernardino. Outdoor Market. National Orange Showgrounds. PH: 213-560-SHOW ext. 12 or 909-795-1884.
Mar 23 CA, Foster City. Bob's Toy Car Show. Holiday Inn, 1221 Chess Dr. SH: 10am-3pm. Bob Goforth, 4061 E. Castro Valley Blvd., Ste. 224, Castro Valley, CA 94552. PH: 510-889-6676 or FAX: 510-581-0397.
Mar 28 CA, National City. San Diego Toy Show. 1545 Tidelands Ave., Suite "J", I-5 S., take R. on 24th St. Exit. SH: 6pm-9pm, T: 25. Edward Morris, PH: 619-441-8198.
Mar 29 CA, Mission Viejo. Toy Collectibles Show. Capistrano Valley H.S., 26301 Via Escolar. SH: 9am-2pm, T: 80. Stephen Spero, PH: 714-768-8901.
Mar 29 CA, Dixon. Doll & Teddy Bear Show. Mayfair Fairgrounds, Fwy. 80 to Hwy. 113 S., 2.5 mi. SH: 10am-3pm. Lloyd Hogan, 5621 Glencrest Ln., Orangevale, CA 95662. PH: 916-989-9291.
Apr 25-27 CA, San Diego. Prestigious Antiques & Collectibles Show. Scottish Rite Center, 1895 Camino Del Rio South. SH: Fri. 12noon-9pm, Sat. 12noon-7pm, Sun. 12noon-5pm. Sy Miller Productions, PO Box 967, Rancho Santa Fe, CA 92067. PH: 619-756-3275.
Apr 25 CA, National City. San Diego Toy Show. 1545 Tidelands Ave., Suite "J", I-5 S., take R. on 24th St. Exit. SH: 6pm-9pm, T: 25. Edward Morris, PH: 619-441-8198.
Apr 26 CA, Mission Viejo. Toy Collectibles Show. Capistrano Valley H.S., 26301 Via Escolar. SH: 9am-2pm, T: 80. Stephen Spero, PH: 714-768-8901.
Apr 27 CA, Foster City. Bob's West Coast Slot-O-Rama. Holiday Inn, 1221 Chess Dr. SH: 10am-3pm. Bob Goforth, 4061 E. Castro Valley Blvd., Ste. 224, Castro Valley, CA 94552. PH: 510-889-6676 or FAX: 510-581-0397.
May 3 CA, San Mateo. Doll & Teddy Bear Show. Cty. Expo Ctr. SH: 10am-3:30pm. Lloyd Hogan, 5621 Glencrest Ln., Orangevale, CA 95662. PH: 916-989-9291.
May 4 CA, Fullerton. Trouble Shooters Antique Round-Up. Campus of Cal. State, off 57 Fwy., 5 min. N. of 91 Fwy. PH: 213-560-SHOW ext. 15.
May 4 CA, Hayward. Antique Toy Show. Centennial Hall, 22292 Foothill Blvd. SH: 10am-3pm, T: 200. Wayne Gateley, PO Box 221052, Sacramento, CA 95822. PH: 916-448-2655.
May 18 CA, Ventura. Flea Market & Swap Meet. Cty. Fairgrounds, Seaside Park. PH: 213-560-SHOW ext. 13.
May 18 CA, Foster City. Bob's Toy Car Show. Holiday Inn, 1221 Chess Dr. SH: 10am-3pm. Bob Goforth, 4061 E. Castro Valley Blvd., Ste. 224, Castro Valley, CA 94552. PH: 510-889-6676 or FAX: 510-581-0397.
May 22-26 CA, San Bernardino. National Orange Show. Nat'l. Orange Showgrounds. PH: 909-888-6788 ext. 431.
May 23 CA, National City. San Diego Toy Show. 1545 Tidelands Ave., Suite "J", I-5 S., take R. on 24th St. Exit. SH: 6pm-9pm, T: 25. Edward Morris, PH: 619-441-8198.
May 31 CA, Mission Viejo. Toy Collectibles Show. Capistrano Valley H.S., 26301 Via Escolar. SH: 9am-2pm, T: 80. Stephen Spero, PH: 714-768-8901.
Jun 7 CA, San Jose. Summer Doll Show. Santa Clara Cty. Fairgrounds, Hwy. 101 to Tully Rd. Exit West. SH: 10am-4pm. Golden Gate Shows, Fern Loiacono, PO Box 1208, Ross, CA 94957. PH: 415-459-1998 or FAX: 415-459-0827.
Jun 21 CA, Santa Rosa. Summer Doll Show. Sonoma Cty. Fairgrounds, Hwy. 101 to Hwy. 12 East, to first offramp (South E St.). SH: 10am-4pm. Golden Gate Shows, Fern Loiacono, PO Box 1208, Ross, CA 94957. PH: 415-459-1998 or FAX: 415-459-0827.
Jun 27 CA, National City. San Diego Toy Show. 1545 Tidelands Ave., Suite "J", I-5 S., take R. on 24th St. Exit. SH: 6pm-9pm, T: 25. Edward Morris, PH: 619-441-8198.
Jun 28 CA, Mission Viejo. Toy Collectibles Show. Capistrano Valley H.S., 26301 Via Escolar. SH: 9am-2pm, T: 80. Stephen Spero, PH: 714-768-8901.
Jul 6 CA, Foster City. Bob's Toy Car Show. Holiday Inn, 1221 Chess Dr. SH: 10am-3pm. Bob Goforth, 4061 E. Castro Valley Blvd., Ste. 224, Castro Valley, CA 94552. PH: 510-889-6676 or FAX: 510-581-0397.
Jul 12-13 CA, Ventura. Doll, Bears, Supplies & Crafts Show. Fairgrounds, 10 W. Harbor Blvd. #101. SH: Sat. 10am-5pm, Sun. 11am-4pm. Miller Production Group, PO Box 967, Rancho Santa Fe, CA 92067. PH: 619-756-3275.
Jul 20 CA, Ventura. Flea Market & Swap Meet. Cty. Fairgrounds, Seaside Park. PH: 213-560-SHOW ext. 13.
Jul 25-27 CA, San Diego. Prestigious Antiques & Collectibles Show. Scottish Rite Center, 1895 Camino Del Rio South. SH: Fri. 12noon-9pm, Sat. 12noon-7pm, Sun. 12noon-5pm. Sy Miller Productions, PO Box 967, Rancho Santa Fe, CA 92067. PH: 619-756-3275.
Jul 25 CA, National City. San Diego Toy Show. 1545 Tidelands Ave., Suite "J", I-5 S., take R. on 24th St. Exit. SH: 6pm-9pm, T: 25. Edward Morris, PH: 619-441-8198.
Aug 16 CA, San Rafael. Summer Doll Show. Marin Cty. Civic Center Exhibition Hall, Hwy. 101 to N. San Pedro Rd. Exit East. SH: 10am-4pm. Golden Gate Shows, Fern Loiacono, PO Box 1208, Ross, CA 94957. PH: 415-459-1998 or FAX: 415-459-0827.
Aug 16-17 CA, Costa Mesa. Dolls, Bears, Supplies & Crafts Show. Orange Co. Fair & Exposition Ctr., Bldg., #12, Arlington Ave. SH: Sat. 10am-5pm, Sun. 11am-4pm. Miller Production Group, PO Box 967, Rancho Santa Fe, CA 92067. PH: 619-756-3275.
Aug 22 CA, National City. San Diego Toy Show. 1545 Tidelands Ave., Suite "J", I-5 S., take R. on 24th St. Exit. SH: 6pm-9pm, T: 25. Edward Morris, PH: 619-441-8198.
Sep 6 CA, Santa Rosa. Fall Doll Show. Sonoma Cty. Fairgrounds, Hwy. 101 to Hwy. 12 East, to first offramp (South E St.). SH: 10am-4pm. Golden Gate Shows, Fern Loiacono, PO Box 1208, Ross, CA 94957. PH: 415-459-1998 or FAX: 415-459-0827.
Sep 7 CA, San Mateo. Toy Show. County Expo Ctr., 2495 S. Delaware St. SH: 10am-3pm, T: 200. Dugger-Hill Promos., PO Box 243, Auburn, CA 95604. PH: 916-888-0291 or 408-476-8612.
Sep 13 CA, San Jose. Fall Doll Show. Santa Clara Cty. Fairgrounds, Hwy. 101 to Tully Rd. Exit West. SH: 10am-4pm. Golden Gate Shows, Fern Loiacono, PO Box 1208, Ross, CA 94957. PH: 415-459-1998 or FAX: 415-459-0827.
Sep 14 CA, Foster City. Bob's Toy Car Show. Holiday Inn, 1221 Chess Dr. SH: 10am-3pm. Bob Goforth, 4061 E. Castro Valley Blvd., Ste. 224, Castro Valley, CA 94552. PH: 510-889-6676 or FAX: 510-581-0397.
Sep 26 CA, National City. San Diego Toy Show. 1545 Tidelands Ave., Suite "J", I-5 S., take R. on 24th St. Exit. SH: 6pm-9pm, T: 25. Edward Morris, PH: 619-441-8198.
Sep 28 CA, Ventura. Flea Market & Swap Meet. Cty. Fairgrounds, Seaside Park. PH: 213-560-SHOW ext. 13.
Oct 5 CA, Fullerton. Trouble Shooters Antique Round-Up. Campus of Cal. State, off 57 Fwy., 5 min. N. of 91 Fwy. PH: 213-560-SHOW ext. 15.
Oct 24 CA, National City. San Diego Toy Show. 1545 Tidelands Ave., Suite "J", I-5 S., take R. on 24th St. Exit. SH: 6pm-9pm, T: 25. Edward Morris, PH: 619-441-8198.
Oct 26 CA, Foster City. Bob's West Coast Slot-O-Rama. Holiday Inn, 1221 Chess Dr. SH: 10am-3pm. Bob Goforth, 4061 E. Castro Valley Blvd., Ste. 224, Castro Valley, CA 94552. PH: 510-889-6676 or FAX: 510-581-0397.

Nov 2 CA, Ventura. Flea Market & Swap Meet. Cty. Fairgrounds, Seaside Park. PH: 213-560-SHOW ext. 13.

Nov 8 CA, San Rafael. Holiday Doll Show. Marin Cty. Civic Center Exhibition Hall, Hwy. 101 to N. San Pedro Rd. Exit East. SH: 10am-4pm. Golden Gate Shows, Fern Loiacono, PO Box 1208, Ross, CA 94957. PH: 415-459-1998 or FAX: 415-459-0827.

Nov 16 CA, Foster City. Bob's Toy Car Show. Holiday Inn, 1221 Chess Dr. SH: 10am-3pm. Bob Goforth, 4061 E. Castro Valley Blvd., Ste. 224, Castro Valley, CA 94552. PH: 510-889-6676 or FAX: 510-581-0397.

Nov 22-23 CA, Ventura. Doll, Bears, Supplies & Crafts Show. Fairgrounds, 10 W. Harbor Blvd. #101. SH: Sat. 10am-5pm, Sun. 11am-4pm. Miller Production Group, PO Box 967, Rancho Santa Fe, CA 92067. PH: 619-756-3275.

Nov 28 CA, National City. San Diego Toy Show. 1545 Tidelands Ave., Suite "J", I-5 S., take R. on 24th St. Exit. SH: 6pm-9pm, T: 25. Edward Morris, PH: 619-441-8198.

Dec 6-7 CA, Costa Mesa. Dolls, Bears, Supplies & Crafts Show. Orange Co. Fair & Exposition Ctr., Bldg., #12, Arlington Ave. SH: Sat. 10am-5pm, Sun. 11am-4pm. Miller Production Group, PO Box 967, Rancho Santa Fe, CA 92067. PH: 619-756-3275.

Dec 26 CA, National City. San Diego Toy Show. 1545 Tidelands Ave., Suite "J", I-5 S., take R. on 24th St. Exit. SH: 6pm-9pm, T: 25. Edward Morris, PH: 619-441-8198.

COLORADO

Feb 22 CO, Loveland. Timber Dan Spring Toy Show. Larimer Cty. Fairgrounds. Timber Dan, 409 Buffalo St., Delano, MN 55328. PH: 612-972-6726.

Apr 13 CO, Denver. "Barbie Madness Mega Show". Marriott South East, 6363 E. Hampden. SH: 10am-4pm. PH: 303-758-7000.

Sep 28 CO, Denver. "Barbie Madness Mega Show". Marriott South East, 6363 E. Hampden. SH: 10am-4pm. PH: 303-758-7000.

CONNECTICUT

Jan 5 CT, Stratford. NY, NH & Hartford Toy & Train Show. Ramada Inn, 225 Lord Ship Blvd. SH: 9am-3pm. Mike Martino, PO Box 9131, New Haven, CT 06532. PH: 203-469-9048.

Jan 12 CT, Waterbury. LLC Winter Model Railroad Show. Four Points Hotel Grand Ballroom, 3580 E. Main St., Exit 25A on 84 E., Exit 26 on 84 W. SH: 9am-3pm. Classic Shows, Ludwig Spinelli, PO Box 2415, Shelton, CT 06484. PH: 203-926-1327.

Jan 26 CT, Wallingford. NY, NH & Hartford Toy & Train Show. Stillwood Inn, 1074 S. Colony Rd., Rt. 5. SH: 9am-2pm. Mike Martino, PO Box 9131, New Haven, CT 06532. PH: 203-469-9048.

Feb 2 CT, Waterbury. Toy Collectible Show. Sheraton Four Point Hotel, 3580 E. Main St., I-84 East Exit 25A, I-84 West Exit 26. SH: 9am-3pm, T: 100-8'. Bernie, PH: 203-274-9592 or 800-595-6697.

Feb 16 CT, Wallingford. NY, NH & Hartford Toy & Train Show. Stillwood Inn, 1074 S. Colony Rd., Rt. 5. SH: 9am-2pm. Mike Martino, PO Box 9131, New Haven, CT 06532. PH: 203-469-9048.

Mar 8-9 CT, E. Hartford. Greenberg's Great Train, Dollhouse & Toy Show. Aero Center, Exit 58 off I-84. SH: Sat. 11am-5pm, Sun. 11am-4pm. Greenberg Shows, 7566 Main St., Sykesville, MD 21784. PH: 410-795-7447.

Mar 23 CT, Trumbull. NY, NH & Hartford Toy & Train Show. Marriott, Hawley Ln. SH: 9am-3pm. Mike Martino, PO Box 9131, New Haven, CT 06532. PH: 203-469-9048.

Apr 27 CT, Wallingford. NY, NH & Hartford Toy & Train Show. Stillwood Inn, 1074 S. Colony Rd., Rt. 5. SH: 9am-2pm. Mike Martino, PO Box 9131, New Haven, CT 06532. PH: 203-469-9048.

May 3 CT, Kent. Carousel Doll Club Antique Collectible Show. Community House, Rt. 7. SH: 10am-3pm, T: 72. Dee Domroe, 9 Stuart Rd., Bridgewater, CT 06752. PH: 860-354-8442.

DELAWARE

Apr 20 DE, Newark. Toys, Comic Books & All Childhood Collectibles. Univ. of Delaware Field House, Exit 1 off I-95, Rt. 896 N. 1 mile. SH: 10am-4pm, T: 400-8'. K.C.P. Inc., 809 Beulah Church St., Apollo, PA 15613. PH: 412-697-5510.

FLORIDA

Jan 4 FL, Ft. Lauderdale. "Barbie Goes To Ft. Lauderdale". Airport Hilton, 1870 Grifin Rd., Dania. SH: 10am-4pm. Marl, PH: 941-751-6275 or Joe, PH: 213-953-6490.

Jan 4-5 FL, Tampa. Bay Area Collectible Toy Supershow. Fort Homer Hesterly Armory, 504 N. Howard Ave. SH: Sat. 10am-5pm, Sun. 11am-4pm, T: 200-8'. Jon, PH: 954-942-9929 or Mark, PH: 561-694-7982.

Jan 4-5 FL, Jacksonville. Greenberg's Great Train, Dollhouse & Toy Show. Fairgrounds, across from Coliseum. SH: Sat. 11am-5pm, Sun. 11am-4pm. Greenberg Shows, 7566 Main St., Sykesville, MD 21784. PH: 410-795-7447.

Jan 4 FL, Tallahassee. Toys, Comics, Sports Cards & Collectibles Show. Marzuq Shrine Temple, 1805 N. Monroe (at W. Tharpe St.). SH: 9am-4pm. Charlie Watson, PH: 904-224-8813.

Jan 11 FL, Winter Haven. Racing Collectible Show. Chain O' Lakes Complex, 210 Cypress Gardens Blvd. SH: 10am-4pm, T: 70. Richard Vick, PO Box 946, Mulberry, FL 33860. PH: 941-646-0245.

Jan 12 FL, Sarasota. Festival of Dollhouse Miniatures. Hyatt Hotel. SH: 10am-5pm. Molly Cromwell, 4701 Duncan Dr., Annandale, VA 22003. PH: 703-978-5353.

Jan 18 FL, Winter Haven. "Anything" Collectible Show. Chain O'Lakes Complex, 210 Cypress Gardens Blvd. SH: 10am-4pm, T: 70. Richard Vick, PO Box 946, Mulberry, FL 33860. PH: 941-646-0245.

Jan 19 FL, Daytona Beach. Toys, Comic Books & All Childhood Collectibles Show. Ocean Center Arena, 101 N. Atlantic, on Rt. A1A. SH: 10am-4pm, T: 450-8'. K.C.P. Inc., 809 Beulah Church Rd., Apollo, PA 15613. PH: 412-697-5510.
Jan 25-26 FL, Orlando. Collectors Expo. Fairgrounds, 4603 W. Colonial Dr. SH: 10am-5pm. Classic Shows, Chip Nofal, PO Box 1507, St. Augustine, FL 32085. PH/FAX: 904-928-0666.
Jan 25-26 FL, Orlando. 8th Annual Florida Extravaganza (FX97). Expo Centre, I-4 Exit 40 or 41 (follow signs to Centroplex & Expo Centre). SH: Sat. 10am-5pm, Sun. 10am-4pm, T: 1000. MARZ Prod. Inc., PH: 813-971-8686.
Jan 25-26 FL, Tampa. Greenberg's Great Train, Dollhouse & Toy Show. Expo Park, 4800 US Hwy. 301 N. SH: Sat. 11am-5pm, Sun. 11am-4pm. Greenberg Shows, 7566 Main St., Sykesville, MD 21784. PH: 410-795-7447.
Jan 25 FL, Winter Haven. Doll & Teddy Bear Show. Chain O'Lakes Complex, 210 Cypress Gardens Blvd. SH: 10am-4pm, T: 70. Richard Vick, PO Box 946, Mulberry, FL 33860. PH: 941-646-0245.
Jan 25 FL, Pompano Beach. 22nd Annual Doll, Bear & Toy Show. Recreation Ctr., 1801 NE 6th St. SH: 10am-5pm. Mildred Swaetzman, PH: 954-434-0818.
Jan 31-Feb 1 FL, Daytona Beach. Sportscar Memorabilia, Collectible Literature & Art Show. Northwest Parking area adj. to Daytona Int'l. Speedway Blvd. SH: Fri. 10:30am-9pm, Sat. 8:30am-12noon, T: 50. Bruce Hollander, 901 S. State Road 7, Hollywood, FL 33023. PH: 954-964-8000 or FAX: 954-964-5969.
Feb 1 FL, Ft. Lauderdale. South Florida Antique & Collectible Toy Super Show. Elks Exhibition Center, 700 NE 10th St., Pompano Beach. SH: 10am-5pm, T: 160-8'. Jon, PH: 954-942-9929 or Mark, PH: 561-694-7982.
Feb 2 FL, Miami. Antique Toy, Doll & Collectible Show. Radisson Mart Plaza, 711 NW 72nd Ave. (opposite the Airport at 72nd Ave., Exit SR 836-Dolphin Expwy). SH: 10am-4pm. PH: 305-446-4488 or 445-2522.
Feb 8-9 FL, Auburndale-Lakeland. Central Florida Toy Expo. Polk County Fairgrounds-Auburndale Speedway, Hwy. 542. SH: 8am-5pm, T: 850. Doug Malcolm, 1420 N. Galloway Rd., Lakeland, FL 33809. PH: 941-686-8320 or Bill & Pat Poe, PH: 904-897-4163.
Feb 8-9 FL, Auburndale-Lakeland. Fast Food Toy Show. Polk Cty. Fairgrounds, Speedway. SH: 8am-4pm, T: 50. Bill Poe, 220 Dominica Circle E., Niceville, FL 32578. PH: 904-897-4163.
Feb 9 FL, West Palm Beach. South Florida Int'l Toy Soldier Show. Airport Holiday Inn, 1301 Belvidere Rd. Tony Angelora, PH: 407-790-4765.
Feb 15 FL, Palm Beach Gardens. Collectible Miniature Show. MacArthur's Holiday Inn. SH: 10am-4:30pm. Bright Star Promos., Inc., 3428 Hillvale Rd., Louisville, KY 40241. PH: 502-423-STAR.
Feb 16 FL, Palm Beach Gardens. Teddy Bear Show. MacArthur's Holiday Inn. SH: 10am-4:30pm. Bright Star Promos., Inc., 3428 Hillvale Rd., Louisville, KY 40241. PH: 502-423-STAR.
Feb 21-22 FL, Sarasta. Spring Grand National. Nat'l. Guard Armory, 2890 Ringling Blvd. SH: 9am-3pm, T: 200. John Fleck, 7004 4th Ave. Dr. NW, Bradenton, FL 34209. PH: 941-795-0900.
Feb 23 FL, Ft. Lauderdale. Great Collectible Toy, Doll, Comic, Model Extravaganza. War Memorial Auditorium, 800 NE 8th St. SH: 10am-5pm, T: 300. Tate's Comics, Tony Ottati, PH: 954-748-0181.
Feb 28-Mar 2 FL, Lakeland. Collectorama Show. Civic Center, 700 W. Lemon. SH: Fri. & Sat. 10am-7pm, Sun. 10am-4pm, T: 200-225 8'. Ed Kuszmar, PO Box 4049, Boca Raton, FL 33429. PH: 561-995-7985. FAX: 561-995-7983.
Mar 1 FL, Kissimmee. 3rd Semi-Annual Doll & Bear Show. Fairgrounds, Valley Agricultural Ctr., 1901 E. Irlo Bronson Memorial Hwy. SH: 10am-5pm, T: 50. Steve Schroeder, 3100 Harvest Ln., Kissimmee, FL 34744. PH: 407-957-6392 or FAX: 407-957-9427.
Mar 1 FL, Cape Coral. 15th Annual Doll, Bear & Minature Show. American Legion Post 90, 923 SE 47th Terr. SH: 10am-4pm. Beverly, PH: 941-542-7253 or Georgia, PH: 941-549-2459.
Mar 8-9 FL, Miami. Antique & Collectible Toy Super Show. MAHI Temple, 1480 NW North River Dr. SH: Sat. 10am-5pm, Sun. 11am-4pm, T: 250-8'. Jon Jacobus, PH: 954-942-9929 or Mark Leinberger, PH: 561-694-7982.
Mar 15-16 FL, Orlando. Antique & Collectible Toy Super Show. Bahia Shrine Auditorium, 2300 Pembrook Dr. SH: Sat. 10am-5pm, Sun. 11am-4pm, T: 250-8'. Jon, PH: 954-942-9929 or Mark, PH: 561-694-7982.
Mar 15 FL, Jacksonville. "River City Doll & Bear Show". Morocco Temple, St. Johns Bluff Rd. SH: 10am-4pm. M. Hill, 13030 Bent Pine Ct. E., Jacksonville, FL 32246. PH: 904-221-1235.
Mar 22 FL, Miami. Teddy Bear Show. Ramada Inn, Exit 103 St. & (826) Palmetto Expy. SH: 10am-4pm. Olga Rodriguez, PH: 305-696-1334.
Mar 29 FL, Ft. Lauderdale. South Florida Antique & Collectible Toy Super Show. Elks Exhibition Center, 700 NE 10th St., Pompano Beach. SH: 10am-5pm, T: 160-8'. Jon, PH: 954-942-9929 or Mark, PH: 561-694-7982.
Apr 5-6 FL, Tampa. Bay Area Collectible Toy Supershow. Fort Homer Hesterly Armory, 504 N. Howard Ave. SH: Sat. 10am-5pm, Sun. 11am-4pm, T: 200-8'. Jon, PH: 954-942-9929 or Mark, PH: 561-694-7982.
Apr 5 FL, Winter Haven. "Anything" Collectible Show. Chain O'Lakes Complex, 210 Cypress Gardens Blvd. SH: 10am-4pm, T: 70. Richard Vick, PO Box 946, Mulberry, FL 33860. PH: 941-646-0245.
Apr 13 FL, Port Richey. Tampa Bay Diecast Toy Show. SH: 10am-4pm, T: 50. Joe Madonia, 12101 Chuck Cir., Hudson, FL 34669. PH: 813-863-7033.
Apr 19 FL, Winter Haven. Racing Collectible Show. Chain O' Lakes Complex, 210 Cypress Gardens Blvd. SH: 10am-4pm, T: 70. Richard Vick, PO Box 946, Mulberry, FL 33860. PH: 941-646-0245.
Apr 19-20 FL, Sarasota. 1st Annual Dolls, Bears, Barbies & More Show. Municipal Auditorium (formerly Exhibition Hall), 801 N. Tamiami Trl., US Hwy. 41. SH: Sat. 10am-5pm, Sun. 11am-4pm. Steve Schroeder, 3100 Harvest Ln., Kissimmee, FL 34744. PH: 407-957-6392 or FAX: 407-957-9427.
May 10-11 FL, Miami. Antique & Collectible Toy Super Show. MAHI Temple, 1480 NW North River Dr. SH: Sat. 10am-5pm, Sun. 11am-4pm, T: 250-8'. Jon Jacobus, PH: 954-942-9929 or Mark Leinberger, PH: 561-694-7982.
May 31 FL, Ft. Lauderdale. South Florida Antique & Collectible Toy Super Show. Elks Exhibition Center, 700 NE 10th St., Pompano Beach. SH: 10am-5pm, T: 160-8'. Jon, PH: 954-942-9929 or Mark, PH: 561-694-7982.
Jun 7-8 FL, Orlando. Antique & Collectible Toy Super Show. Bahia Shrine Auditorium, 2300 Pembrook Dr. SH: Sat. 10am-5pm, Sun. 11am-4pm, T: 250-8'. Jon, PH: 954-942-9929 or Mark, PH: 561-694-7982.
Jun 7 FL, Winter Haven. Doll & Teddy Bear Show. Chain O' Lakes Complex, 210 Cypress Gardens Blvd. SH: 10am-4pm, T: 70. Richard Vick, PO Box 946, Mulberry, FL 33860. PH: 941-646-0245.
Jul 18-20 FL, Lakeland. Collectorama Show. Civic Center, 700 W. Lemon. SH: Fri. & Sat. 10am-7pm, Sun. 10am-4pm, T: 200-225 8'. Ed Kuszmar, PO Box 4049, Boca Raton, FL 33429. PH: 561-995-7985. FAX: 561-995-7983.
Aug 16-17 FL, Orlando. Antique & Collectible Toy Super Show. Bahia Shrine Auditorium, 2300 Pembrook Dr. SH: Sat. 10am-5pm, Sun. 11am-4pm, T: 250-8'. Jon, PH: 954-942-9929 or Mark, PH: 561-694-7982.
Aug 23-24 FL, Miami. Antique & Collectible Toy Super Show. MAHI Temple, 1480 NW North River Dr. SH: Sat. 10am-5pm, Sun. 11am-4pm, T: 250-8'. Jon Jacobus, PH: 954-942-9929 or Mark Leinberger, PH: 561-694-7982.
Aug 23 FL, Winter Haven. Racing Collectible Show. Chain O' Lakes Complex, 210 Cypress Gardens Blvd. SH: 10am-4pm, T: 90. Richard Vick, PO Box 946, Mulberry, FL 33860. PH: 941-646-0245.
Oct 4 FL, Winter Haven. Doll & Teddy Bear Show. Chain O' Lakes Complex, 210 Cypress Gardens Blvd. SH: 10am-4pm, T: 70. Richard Vick, PO Box 946, Mulberry, FL 33860. PH: 941-646-0245.
Oct 17-19 FL, Lakeland. Collectorama Show. Civic Center, 700 W. Lemon. SH: Fri. & Sat. 10am-7pm, Sun. 10am-4pm, T: 200-225 8'. Ed Kuszmar, PO Box 4049, Boca Raton, FL 33429. PH: 561-995-7985. FAX: 561-995-7983.
Nov 8 FL, Winter Haven. "Anything" Collectible Show. Chain O' Lakes Complex, 210 Cypress Gardens Blvd. SH: 10am-4pm, T: 70. Richard Vick, PO Box 946, Mulberry, FL 33860. PH: 941-646-0245.

GEORGIA

Jan 5 GA, Atlanta. Antique Toy, Doll, Train & Collectible Show. Cobb Cty. Civic Ctr. (Marietta), I-75, Exit 112, W. 2 mi. SH: 10am-3pm, T: 300. DeSalle Promos., Inc., 5106 Knollwood Ln., Anderson, IN 46011. PH: 800-392-TOYS.
Jan 11-12 GA, Norcross. Greenberg's Great Train, Dollhouse & Toy Show. North Atlanta Trade Ctr., 5 mi. N. of Atlanta Perimeter Hwy. SH: Sat. 11am-5pm, Sun. 11am-4pm. Greenberg Shows, 7566 Main St., Sykesville, MD 21784. PH: 410-795-7447.
Mar 2 GA, Atlanta. "Barbie Goes To Atlanta". Renaissance Hotel-Atlanta Airport, 4736 Best Rd. SH: 10am-4pm. Marl, PH: 941-751-6275 or Joe, PH: 213-953-6490.
Apr 22-26 GA, Cleveland. 10th Annual Cabbage Patch Kids Collector's Club Convention. Home of the original Cabbage Patch Kids. Original Appalachian Artworks, Inc., PH: 706-865-2171.

HAWAII

Jan 23-30 HI, Honolulu. Doll Festival. Regent Hotel. Rowbear Lowman, PO Box 66823, Scotts Valley, CA 95067. PH: 408-438-5349.
Jan 30-Feb 6 HI, Honolulu. Hawaiian Doll Festival & Convention. Regent Hotel. Rowbear Lowman, PO Box 66823, Scotts Valley, CA 95067. PH: 408-438-5349.
Feb 8 HI, Honolulu. Barbie Club of Hawaii Mini-Convention. Prince Hotel, 100 Holomoana St. SH: 10am-3pm. Connie Holladay, PO Box 574, Kaneohe, HI 96744. PH: 808-247-6585.

IDAHO

Apr 12-13 ID, Boise. Toy Show. Western ID Fairgrounds. Kevin or Barbara, PH: 208-887-4847 or 890-5255.

ILLINOIS

Jan 11-12 IL, Decatur. 12th Annual Farm Toy Show & AUCTION. Holiday Inn, US 36 W. & Wykles Rd. SH: Sat. auction 10am, Sun. show 9am-3pm. Dary & Diana Burnett, Rt. 3, Box 188, Decatur, IL 62526. PH: 217-963-2988 or Calvin & Sandee Elder, PH: 217-668-2436.
Jan 12 IL, Bridgeview. "If Its Got Wheels" Swap Meet. John A. Oremus Community Center, 7900 S. Oketo Ave. SH: 8am-3pm, T: 8'. Jim LaCoco, 231 Englewood Ave., Bellwood, IL 60104. PH: 708-544-1975.
Jan 19 IL, Bridgeview. Comic & Action Figure Show. Oremus Ctr., 79th & Oketo. SH: 10am-4pm, T: 55. Barry Huebner, 10144 S. Roberts Rd., Palos Hills, IL 60465. PH: 708-233-0037.
Jan 26 IL, Naperville. Antique & Collectible Doll & Bear Show. Holiday Inn, 1801 Naperville Rd. SH: 9am-4pm. Gould & Gallup, 28 W. 491 87th St., Naperville, IL 60564. FAX: 630-355-0574.

Feb 2 IL, Joliet. Toyfest IV. Holiday Inn, I-80 & Larkin Ave. SH: 9am-3pm, T: 60. Edward Streich, 24728 W. Hemphill Dr., Elwood, IL 60426. PH: 815-478-4362.
Feb 9 IL, Westmont. Antique, Collectible, Modern Doll, Teddy Bear & Miniature Show. Inland Meeting & Exposition Center, 400 E. Ogden, I-294 (Rt. 34), 3 mi. W. of 294 (34) on North Side. SH: 9am-4pm. Gigi's Dolls & Sherry's Teddy Bears, Inc., 6029 N. Northwest Hwy., Chicago, IL 60631. PH: 312-594-1540.
Feb 15-16 IL, Dolton. Santa's Antique & Collectible Toy & Advertising Show. Expo Ctr., 14200 Chicago Rd. SH: Sat. 12noon-5pm, Sun. 9am-2pm. PH: 708-895-5293.
Feb 23 IL, Wheaton. 29th Illinois Plastic Kit & Toy Show. Dupage Cty. Fairgrounds, Exhibition Ctr., 2015 W. Manchester Rd. SH: 9am-3pm, T: 250. Past-Time Hobbies, Inc., George, PH: 630-969-1847 or 708-993-0883 contest info.
Feb 23 IL, Elgin. Auto Racing & Collectible Show. Holiday Inn, 345 River Rd., 1 blk. S. of I-90, Rt. 31 Exit. SH: 10am-3pm, T: 60. Doug Toussaint, PO Box 53, Ringwood, IL 60072. PH: 815-675-6017.
Mar 1 IL, Schiller Park-Chicago. 7th Annual Strictly 43rd Model Expo & Contest (Swap Meet). Ballroom of the Howard Johnson O'Hare International, Manheim & Irving Park Rd. SH: 9am-3:30pm. Don Anderson, PH: 630-964-6101.
Mar 2 IL, Rockford. Doll, Toy & Bear Show. Forest Hills Lodge, 9900 Forest Hills Rd. & Rt. 173. SH: 9am-4pm, T: 160. JoAnn Reynolds, 6058 S. Daysville Rd., Oregon, IL 61061. PH: 815-732-7742.
Mar 2 IL, Grayslake. Nostalgia Toy & Doll Show. Lake Co. Fairgrounds, Rt. 120 & Rt. 45. SH: 9am-4pm, T: 975. Skip's, Joe Beelel, PO Box 88266, Carol Stream, IL 60188. PH: 630-682-8792 or 800-250-7369.
Mar 23 IL, Gurnee. Teddy Bear Show. Holiday Inn, 6161 W. Grand Ave. SH: 9:30am-4pm, T: 60-70. Marge Hansen, N96W20235 County Line Rd., Meno. Falls, WI 53051. PH: 414-255-4465.
Mar 23 IL, Elgin. Auto Racing & Collectible Show. Holiday Inn, 345 River Rd., 1 blk. S. of I-90, Rt. 31 Exit. SH: 10am-3pm, T: 60. Doug Toussaint, PO Box 53, Ringwood, IL 60072. PH: 815-675-6017.
Apr 6 IL, Bridgeview. Comic & Action Figure Show. Oremus Ctr., 79th & Oketo. SH: 10am-4pm, T: 55. Barry Huebner, 10144 S. Roberts Rd., Palos Hills, IL 60465. PH: 708-233-0037.
Apr 6 IL, Chicago. "Barbie Madness Mega Show". Schaumburg Marriott, 50 N. Martingale Rd. SH: 10am-4pm. PH: 847-240-0100.
Apr 13 IL, LaSalle-Peru. Doll, Toy & Bear Show. Illinois Valley Banquet Ctr., 920 2nd St. SH: 9am-4pm, T: 150. JoAnn Reynolds, 6058 S. Daysville Rd., Oregon, IL 61061. PH: 815-732-7742.
Apr 13 IL, Bloomington. Antique & Collectible Doll, Bear, Toy & Miniature Show. Interstate Ctr., 2301 W. Market St. (Rt. 9). SH: 9am-4pm. Gould & Gallup, 28 W. 491 87th St., Naperville, IL 60564. FAX: 630-355-0574.
Apr 20 IL, Elgin. Auto Racing & Collectible Show. Holiday Inn, 345 River Rd., 1 blk. S. of I-90, Rt. 31 Exit. SH: 10am-3pm, T: 60. Doug Toussaint, PO Box 53, Ringwood, IL 60072. PH: 815-675-6017.
Apr 27 IL, Saint Charles. Antique Toy & Doll World Shows. Kane Cty. Fairgrounds, Rt. 64 & Randall Rd. SH: 8:30am-4pm. Antique World Shows, PO Box 34509, Chicago, IL 60634. PH: 847-526-1645.
May 3-4 IL, Hillside-Chicago. Reznhedz Model & Toy Show. Holiday Inn, 4400 Frontage Rd. SH: Sat. 10am-6pm, Sun. 11am-5pm, T: 100. Larry Burbridge, 3373 B Beacon #6, N. Chicago, IL 60064. PH: 847-473-1821 after 6pm CST.
May 25 IL, Grayslake. Nostalgia Toy & Doll Show. Lake Co. Fairgrounds, Rt. 120 & Rt. 45. SH: 9am-4pm, T: 975. Skip', Joe Beelel, PO Box 88266, Carol Stream, IL 60188. PH: 630-682-8792 or 800-250-7369.
Jun 22 IL, Saint Charles. Antique Toy & Doll World Shows. Kane Cty. Fairgrounds, Rt. 64 & Randall Rd. SH: 8:30am-4pm. Antique World Shows, PO Box 34509, Chicago, IL 60634. PH: 847-526-1645.
Jun 22 IL, Bridgeview. Comic & Action Figure Show. Oremus Ctr., 79th & Oketo. SH: 10am-4pm, T: 55. Barry Huebner, 10144 S. Roberts Rd., Palos Hills, IL 60465. PH: 708-233-0037.
Jul 20 IL, Naperville. Antique & Collectible Doll & Bear Show. Holiday Inn, 1801 Naperville Rd. SH: 9am-4pm. Gould & Gallup, 28 W. 491 87th St., Naperville, IL 60564. FAX: 630-355-0574.
Aug 10 IL, Rockford. Doll, Toy & Bear Show. Forest Hills Lodge, 9900 Forest Hills Rd. & Rt. 173. SH: 9am-4pm, T: 160. JoAnn Reynolds, 6058 S. Daysville Rd., Oregon, IL 61061. PH: 815-732-7742.
Sep 28 IL, Westmont. Antique, Collectible, Modern Doll, Teddy Bear & Miniature Show. Inland Meeting & Exposition Center, 400 E. Ogden, I-294 (Rt. 34), 3 mi. W. of 294 (34) on North Side. SH: 9am-4pm. Gigi's Dolls & Sherry's Teddy Bears, Inc., 6029 N. Northwest Hwy., Chicago, IL 60631. PH: 312-594-1540.
Oct 12 IL, La Salle-Peru. Doll, Toy & Bear Show. Illinois Valley Banquet Ctr., 920 2nd St. SH: 9am-4pm, T: 150. JoAnn Reynolds, 6058 Daysville Rd., Oregon, IL 61061. PH: 815-732-7742.
Oct 12 IL, Bloomington. Antique & Collectible Doll, Bear, Toy & Miniature Show. Interstate Ctr., 2301 W. Market St. (Rt. 9). SH: 9am-4pm. Gould & Gallup, 28 W. 491 87th St., Naperville, IL 60564. FAX: 630-355-0574.
Oct 26 IL, Saint Charles. Antique Toy & Doll World Shows. Kane Cty. Fairgrounds, Rt. 64 & Randall Rd. SH: 8:30am-4pm. Antique World Shows, PO Box 34509, Chicago, IL 60634. PH: 847-526-1645.
Nov 13-16 IL, Rosemont. Pinball Expo-Flip Out Pinball Tournament. Ramada Hotel, 6600 N. Mannheim Rd. SH: Thurs.-Sun. 9am-12midnight, T: 80. Robert Berk, 2671 Youngstown Rd. SE, Warren, OH 44484. PH: 330-369-1192 or 800-323-3547.

INDIANA

Mar 9 IN, Ft. Wayne. Doll & Bear Show. The Lantern, 4420 Ardmore Ave. SH: 9:30am-3:30pm. B&L Promos., PH: 419-228-4657.

Mar 23 IN, Evansville. Collectible Miniature Show. Marriott Airport. SH: 10am-4:30pm. Bright Star Promos., Inc., 3428 Hillvale Rd., Louisville, KY 40241. PH: 502-423-STAR.
Jun 1 IN, Indianapolis. Teddy Bear Show. Ritz Charles Center, 12156 US 31 N. of I-465. SH: 10am-4:30pm. Bright Star Promos., Inc., 3428 Hillvale Rd., Louisville, KY 40241. PH: 502-423-STAR.
Jun 1 IN, Indianapolis. "Barbie Madness Mega Show". Ramada Inn, 7701 E. 42nd St. SH: 10am-4pm. PH: 317-897-4000.
Jun 8 IN, Indianapolis. Collectible Miniature Show. Holiday Inn North. SH: 10am-4:30pm. Bright Star Promos., Inc., 3428 Hillvale Rd., Louisville, KY 40241. PH: 502-423-STAR.
Sep 1 IN, Goshen. Diecast Toy Show. Elkhart Cty. 4-H Fairgrounds, E. Monroe St. SH: 8:30am-3pm, T: 275. Ken Miller, 1701 Berkey Ave., Goshen, IN 46526. PH: 219-534-2719.

IOWA

Jan 19 IA, Central City. 5th Annual Train Show & Swap Meet. High School. SH: 10am-4pm, T: 120. Donald Earll, 201 10th Ave., Hiawatha, IA 52233. PH: 319-393-2691.
Mar 9 IA, Maquoketa. 15th Annual Doll, Toy & Bear Show. Jackson Cty. Fairgrounds, Jct. Hwys. 62 & 64. SH: 9am-4pm. Dora Pitts, 4697 155th St., Clinton, IA 52732. PH: 319-242-0139.
Mar 9 IA, Calamus. 9th Annual Farm Toy Show. C/W Elementary School, S. end of 2nd St. SH: 9am-3pm. "Yogi" Braet, 308 Clinton St., Calamus, IA 52729. PH: 319-246-2668 or Norm Bousselot, PH: 319-246-2628.
Mar 22 IA, Davenport. Quad Cities Collectables Paradise. Mississippi Valley Fairgrounds, 2815 W. Locust. SH: 10am-4pm, T: 80. Vince Jones, PO Box 172, Davenport, IA 52805. PH: 319-386-5763.
Mar 22 IA, Des Moines. National Transportation Toy Show. Tourism Bldg., State Fairgrounds, E. 30th & University Ave. SH: 9am-4pm, T: 120. Marge Lowry, 1050 24th St., Des Moines, IA 50311. PH: 515-279-9998.
Apr 19-20 IA, Davenport. Quad Cities Collectables Paradise. Mississippi Valley Fairgrounds, 2815 W. Locust. SH: 10am-4pm, T: 80. Vince Jones, PO Box 172, Davenport, IA 52805. PH: 319-386-5763.
May 10 IA, Des Moines. 9th Annual Doll, Toy & Bear Show. Metro Ice Sports Arena, 7201 Hickman Rd. SH: 9am-4pm. Dora Pitts, 4697 155th St., Clinton, IA 52732. PH: 319-242-0139.

KANSAS

Mar 9 KS, Kansas City. 5th Annual Fall Toy & Doll Show. Jack Reardon Civic Center, 5th & Minnesota. SH: 10am-4pm. Judy & Bob Condray, 1005 W. 11th, Concordia, KS 66901. PH: 913-243-3774 or 455-3440.
Mar 16 KS, Wichita. Teddy Bear Show. Marriott Hotel. SH: 10am-4:30pm. Bright Star Promos., Inc., 3428 Hillvale Rd., Louisville, KY 40241. PH: 502-423-STAR.
Sep 7 KS, Kansas City. 18th Annual Fall Toy & Doll Show. Jack Reardon Civic Center, 5th & Minnesota. SH: 10am-4pm. Judy & Bob Condray, 1005 W. 11th, Concordia, KS 66901. PH: 913-243-3774 or 455-3440.

KENTUCKY

Jan 5 KY, Louisville. Collectible Racing & Hot Wheels Expo. Holiday Inn Hurstbourne, I-64 at Hurstbourne Pky. SH: 9am-3pm, T: 32. Redline Racing, 10250 Cedarwood Dr., Union, KY 41091. PH: 606-384-3854.
Feb 9 KY, Ft. Mitchell-Cincinnati, OH. Collectible Miniatures Show. Drawbridge Inn Estates. SH: 10am-4:30pm. Bright Star Promos., Inc., 3428 Hillvale Rd., Louisville, KY 40241. PH: 502-423-STAR.
Apr 6 KY, Louisville. Collectible Miniature Show. Holiday Inn Hurstbourne. SH: 10am-4:30pm. Bright Star Promos., Inc., 3428 Hillvale Rd., Louisville, KY 40241. PH: 502-423-STAR.
Apr 13 KY, Ft. Mitchell-Cincinnati, OH. Teddy Bear Show. Drawbridge Inn Estates. SH: 10am-4:30pm. Bright Star Promos., Inc., 3428 Hillvale Rd., Louisville, KY 40241. PH: 502-423-STAR.

LOUISIANA

Jan 18-19 LA, Baton Rouge. Greenberg's Great Train, Dollhouse & Toy Show. Riverside Centroplex, on banks of Mississippi River. SH: Sat. 11am-5pm, Sun. 11am-4pm. Greenberg Shows, 7566 Main St., Sykesville, MD 21784. PH: 410-795-7447.
Feb 15 LA, Bossier City. 6th Annual River Rouge Doll & Toy Club Show. Parkway High School, 4301 Panther Dr. SH: 9am-4pm, T: 100. Helen Keller, 1807 Pluto Dr., Bossier City, LA 71112. PH: 318-747-3628.

MARYLAND

Jan 25-26 MD, Upper Marlboro. Greenberg's Great Train, Dollhouse & Toy Show. Show Place Arena, Rts. 301 & 4. SH: Sat. 11am-5pm, Sun. 11am-4pm. Greenberg Shows, 7566 Main St., Sykesville, MD 21784. PH: 410-795-7447.
Mar 1-2 MD, Gaithersburg. Eastern National Dolls, Toys & Games Show. Fairgrounds-Agricultural Center, 12 miles NW of Washington DC, Exit 11, off I-270. SH: Sat. 10am-5pm, Sun. 11am-4pm, T: 600. A Bellman Collectors Event, PH: 410-329-2188 or 301-963-3106 (during show hrs.).
Mar 2 MD, Rockville. Mid-Atlantic Dollhouse Miniatures Show. DoubleTree Hotel. Molly Cromwell, 4701 Duncan Dr., Annandale, VA 22003. PH: 703-978-5353.
Mar 15-16 MD, Timonium. Greenberg's Great Train, Dollhouse & Toy Show. Fairgrounds, Exit 16A or 17 off I-83. SH: Sat. 11am-5pm, Sun. 11am-4pm. Greenberg Shows, 7566 Main St., Sykesville, MD 21784. PH: 410-795-7447.
Apr 20 MD, Bel Air. Annual Toy & Train Show. Vol. Fire Co. Main Station, 109 S. Hickory Ave. SH: 9am-3pm, T: 8'. Joe Rutherford, PH: 410-879-9088 or Tom Schaech, PH: 410-838-9436.
May 24-25 MD, Annapolis. GI Joe International Collectors Club 10th Convention. Medford Armory, 18 Willow St. SH: Sat. 10am-5pm, Sun. 9am-3pm, T: 100. James DeSimone, 150 S. Glenoaks Blvd., Burbank, CA 91510. PH: 818-563-1179.
Jun 7-8 MD, Gaithersburg. Eastern National Dolls, Toys & Games Show. Fairgrounds-Agricultural Center, 12 miles NW of Washington DC, Exit 11, off I-270. SH: Sat. 10am-5pm, Sun. 11am-4pm, T: 600. A Bellman Collectors Event, PH: 410-329-2188 or 301-963-3106 (during show hrs.).
Sep 7 MD, Timonium. Baltimore Festival of Dollhouse Miniatures. Holiday Inn. Molly Cromwell, 4701 Duncan Dr., Annandale, VA 22003. PH: 703-978-5353.

MASSACHUSETTS

Jan 5 MA, Lenox. "Philmont" Mountain Model Railroad Club Toy & Train Swap Meet. Memorial High School, East St., Exit 2 Mass Pike approx. 5 mi. N. Rt. 7, R. Walker St. SH: 9am-2pm. Philmont Model Railroad Club, Rick Washburn, 12 Hudson St., Hudson, NY 12534. PH: 518-828-6508.
Jan 19 MA, Auburn. Doll, Bear & Miniature Show. Ramada Inn, 624 Southbridge St., Rt. 12, Exit 10 off Mass Pike. Cindy Amburgey, 47 Jackson Ave., Fitchburg, MA 01420. PH: 508-342-8292 or 342-2265.
Feb 9 MA, Dedham. Cabin Fever Toy Show. Holiday Inn, Jcts. of Rts. 1 & 128-95, Exit 15A. SH: 9am-2:30pm. Atlantic Group, Mrs. Devlin, PO Box 217, Swansea, MA 02777. PH: 508-379-9733.
Feb 16 MA, South Attleboro. Collectable Show. K of C, 305 Highland Ave. SH: 10am-4pm, T: 59. Harry Martin, 18 Vinald Rd., Medfield, MA 02052. PH: 508-359-8611.
Mar 16 MA, Dedham. Greater Boston Antique & Collectible Toy Show. Holiday Inn, Jcts. of Rts. 1 & 128-95, Exit 15A. 100 booths. Atlantic Group, Mrs. Devlin, PO Box 217, Swansea, MA 02777. PH: 508-379-9733.
Mar 16 MA, Marlboro. Trucks, Toys & Udda T'ings. New England Sports Ctr., 121 Donald Lynch Blvd. SH: 9am-2pm, T: 150-8'. Rene Fortin, 116 Gold St., Manchester, NH 03103. PH: 603-622-8649.
Apr 5-6 MA, Wilmington. Greenberg's Great Train, Dollhouse & Toy Show. Shriners Auditorium, I-93, Exit 39. SH: Sat. 11am-5pm, Sun. 11am-4pm. Greenberg Shows, 7566 Main St., Sykesville, MD 21784. PH: 410-795-7447.
Apr 6 MA, Peabody. 8th Annual East Mass Model & Toy Show. Holiday Inn, 1 Newbury St. (Rt. 1 N.). SH: 9am-2pm, T: 60. Michael Zelikson, 990 Paradise Rd., Ste. 3G, Swampscott, MA 01907. PH: 617-631-9677.
Apr 13 MA, Dedham. 14th Annual Spring Show. Holiday Inn, Jct. Rts. 1 & 128 (95). SH: 9am-3pm, T: 100. Northeast Toy Soldier Society, 12 Beach Rd., Gloucester, MA 01930. PH: 508-283-2612.
Apr 13 MA, South Attleboro. Collectable Show. K of C, 305 Highland Ave. SH: 10am-4pm, T: 59. Harry Martin, 18 Vinald Rd., Medfield, MA 02052. PH: 508-359-8611.
Apr 20 MA, Chelmsford. Merrimack Valley Toy Show. Rt. 110, Exit 34 off Rt. 495. SH: 9am-2:30pm, T: 50. Ross Tracy, 7 Parsonage Hill Rd., Haverhill, MA 01832. PH: 508-373-2904.
May 3 MA, Boston. "Barbie Madness Mega Show". Holiday Inn Dedham, 55 Ariadne Rd. SH: 10am-4pm. PH: 617-329-1000 Sat. 9am-3pm.
May 13-18 MA, Brimfield. Antique Show. Central Park. SH: 6am. PO Box 224, Brimfield, MA 01010. PH: 413-596-9257 or 245-4674.
May 18 MA, Auburn. Massachusett's Priemier All Barbie Show. Ramada Inn, 624 Southbridge St. (Rt. 12) (Exit 10 off Mass Pike). SH: 10am-3pm. Cindy Amburgey, 47 Jackson Ave., Fitchburg, MA 01420. PH: 508-342-8292 or 342-2265.
May 18 MA, South Attleboro. Collectable Show. K of C, 305 Highland Ave. SH: 10am-4pm, T: 59. Harry Martin, 18 Vinald Rd., Medfield, MA 02052. PH: 508-359-8611.
Jun 1 MA, Sturbridge. Festival of Dollhouse Miniatures. Host Hotel on Cedar Lake. Molly Cromwell, 4701 Duncan Dr., Annandale, VA 22003. PH: 703-978-5353.
Jun 1 MA, Pittsfield. Collectible & Antique Doll & Toy Show. Berkshire Hilton Inn. SH: 10am-4pm. Jan Fennick, PO Box 9, Commack, NY 11725. PH: 516-266-3190.
Jun 29 MA, South Attleboro. Collectable Show. K of C, 305 Highland Ave. SH: 10am-4pm, T: 59. Harry Martin, 18 Vinald Rd., Medfield, MA 02052. PH: 508-359-8611.
Jul 8-13 MA, Brimfield. Antique Show. Central Park. SH: 6am. PO Box 224, Brimfield, MA 01010. PH: 413-596-9257 or 245-4674.
Aug 24 MA, South Attleboro. Collectable Show. K of C, 305 Highland Ave. SH: 10am-4pm, T: 59. Harry Martin, 18 Vinald Rd., Medfield, MA 02052. PH: 508-359-8611.
Sep 2-7 MA, Brimfield. Antique Show. Central Park. SH: 6am. PO Box 224, Brimfield, MA 01010. PH: 413-596-9257 or 245-4674.
Sep 28 MA, South Attleboro. Collectable Show. K of C, 305 Highland Ave. SH: 10am-4pm, T: 59. Harry Martin, 18 Vinald Rd., Medfield, MA 02052. PH: 508-359-8611.
Sep 28 MA, Dedham. Greater Boston Antique & Collectible Toy Show. Holiday Inn, Jcts. of Rts. 1 & 128-95, Exit 15A. 100 booths. Atlantic Group, Mrs. Devlin, PO Box 217, Swansea, MA 02777. PH: 508-379-9733.
Oct 18 MA, Dedham. 4th Annual Fall Show. Holiday Inn, Jct. Rts. 1 & 128 (95). SH: 9am-3pm, T: 100. Northeast Toy Soldier Society, 12 Beach Rd., Gloucester, MA 01930. PH: 508-283-2612.
Nov 2 MA, Dedham. Greater Boston Antique & Collectible Toy Show. Holiday Inn, Jcts. of Rts. 1 & 128-95, Exit 15A. 100 booths. Atlantic Group, Mrs. Devlin, PO Box 217, Swansea, MA 02777. PH: 508-379-9733.
Nov 2 MA, Milford. Toy Show II. Radisson Hotel, Rt. 109. SH: 9am-2:30pm, T: 75. Ross Tracy, 7 Parsonage Hill Rd., Haverhill, MA 01832. PH: 508-373-2904.
Nov 16 MA, South Attleboro. Collectable Show. K of C, 305 Highland Ave. SH: 10am-4pm, T: 59. Harry Martin, 18 Vinald Rd., Medfield, MA 02052. PH: 508-359-8611.
Dec 21 MA, South Attleboro. Collectable Show. K of C, 305 Highland Ave. SH: 10am-4pm, T: 59. Harry Martin, 18 Vinald Rd., Medfield, MA 02052. PH: 508-359-8611.

MICHIGAN

Jan 1 MI, Eastpointe. New Years Toys, Sports Cards & Music Show. Eagles Hall, on 8 Mile Rd., W. of Gratiot. SH: 10am-4pm, T: 50. Blackjack Promos., Jack, PH: 810-585-0152 7pm-11pm.
Jan 1 MI, St. Heights. Collectable Toy & Modeler Show. Freedom Hill, 15000 16 Mile Rd. SH: 10am-3pm. Joseph Pellegrino, PH: 810-792-9461 or John Rogers, PH: 810-939-4205.
Jan 5-6 MI, Dearborn. Antique Toy Show. Civic Center, 15801 Michigan Ave. SH: 9am-3pm, T: 300. Kevin Hauser or Eric Olbrich, 710 W. 11 Mile Rd., Ste. 121, Royal Oak, MI 48067. PH: 810-399-0155.
Jan 18 MI, Plymouth. Toy & Train Show. Cultural Ctr., 525 Farmer. SH: 11am-3pm. R.R. Promos., Inc., Box 6094, Plymouth, MI 48170. PH: 313-455-2110.
Feb 15-16 MI, Detroit. 6th Annual Toy-A-Rama. Cobo Ctr., 1 Washington Blvd. SH: Sat. 11am-10pm, Sun. 11am-5pm, T: 200. Jerry Patlow, 29276 Geraldine Ct., Farmington Hills, MI 48336. PH: 810-477-0200 or FAX: 810-477-0203.
Mar 1-2 MI, Grand Rapids. 2nd Annual Toy-A-Rama. Grand Ctr., 245 Monroe Ave. NW. SH: Sat. 11am-10pm, Sun. 11am-5pm, T: 200. Jerry Patlow, 29276 Geraldine Ct., Farmington Hills, MI 48336. PH: 810-477-0200 or FAX: 810-477-0203.
Mar 8-9 MI, Novi. Greenberg's Great Train, Dollhouse & Toy Show. Expo Ctr., 43700 Expo Center Dr. SH: Sat. 11am-5pm, Sun. 11am-4pm. Greenberg Shows, 7566 Main St., Sykesville, MD 21784. PH: 410-795-7447.
Mar 22-23 MI, Dearborn. Antique Toy Show. Civic Center, 15801 Michigan Ave. SH: 9am-3pm, T: 300. Kevin Hauser or Eric Olbrich, 710 W. 11 Mile Rd., Ste. 121, Royal Oak, MI 48067. PH: 810-399-0155.
Apr 26 MI, Kalamazoo. Collectible Miniature Show. Cty. Fairgrounds, I-94 Business Loop to Lake St. SH: 10am-4:30pm. Bright Star Promos., Inc., 3428 Hillvale Rd., Louisville, KY 40241. PH: 502-423-STAR.
Apr 27 MI, Kalamazoo. Teddy Bear Show. Cty. Fairgrounds. SH: 10am-4:30pm. Bright Star Promos., Inc., 3428 Hillvale Rd., Louisville, KY 40241. PH: 502-423-STAR.
Apr 27 MI, Livonia. Spring '97 WCCA. Monaghan K of C Hall, 19801 Farmington Rd. SH: 10am-3pm, T: 62. Stan Wutka, 5985 Leland Dr., Ann Arbor, MI 48105. PH: 313-747-7192.
May 10-11 MI, Dearborn. Antique Toy Show. Civic Center, 15801 Michigan Ave. SH: 9am-3pm, T: 300. Kevin Hauser or Eric Olbrich, 710 W. 11 Mile Rd., Ste. 121, Royal Oak, MI 48067. PH: 810-399-0155.
May 31-Jun 1 MI, Midland. Antique Show, Cty. Fairgrounds. Fairgrounds, Eastman Ave. at US 10. SH: Sat. 8am-7pm, Sun. 8am-4pm, T: 1,000. 2156 Rudy Ct., Midland, MI 48642. PH: 517-687-9005 7pm-9pm Mon.-Fri.
Jul 19 MI, Grand Rapids. Collectible Miniatures Show. Hilton Inn Airport, 4747 28th St., just off I-96. SH: 10am-4:30pm. Bright Star Promos., Inc., 3428 Hillvale Rd., Louisville, KY 40241. PH: 502-423-STAR.
Jul 20 MI, Ann Arbor. Teddy Bear Show. Weber's Inn, I-94, Exit 172. SH: 10am-4:30pm. Bright Star Promos., Inc., 3428 Hillvale Rd., Louisville, KY 40241. PH: 502-423-STAR.
Jul 26-27 MI, Midland. Antique Show, Cty. Fairgrounds. Fairgrounds, Eastman Ave. at US 10. SH: Sat. 8am-7pm, Sun. 8am-4pm, T: 1,000. 2156 Rudy Ct., Midland, MI 48642. PH: 517-687-9005 7pm-9pm Mon.-Fri.
Aug 10 MI, Detroit. "Barbie Madness Mega Show". Ypsilanti Marriott, 1275 Huron St. SH: 10am-4pm. PH: 313-758-7000.
Sep 20-21 MI, Dearborn. Antique Toy Show. Civic Center, 15801 Michigan Ave. SH: 9am-3pm, T: 300. Kevin Hauser or Eric Olbrich, 710 W. 11 Mile Rd., Ste. 121, Royal Oak, MI 48067. PH: 810-399-0155.
Sep 27-28 MI, Midland. Antique Show, Cty. Fairgrounds. Fairgrounds, Eastman Ave. at US 10. SH: Sat. 8am-7pm, Sun. 8am-4pm, T: 1,000. 2156 Rudy Ct., Midland, MI 48642. PH: 517-687-9005 7pm-9pm Mon.-Fri.
Oct 19 MI, Livonia. Fall '97 WCCA. Monaghan K of C Hall, 19801 Farmington Rd. SH: 10am-3pm, T: 62. Stan Wutka, 5985 Leland Dr., Ann Arbor, MI 48105. PH: 313-747-7192.

MINNESOTA

Jan 5 MN, Minneapolis. Mid-Winter Toy Show. Northwest Inn Hotel, Hwy. 694 & Hwy. 81 N. SH: 9:30am-2:30pm, T: 130. Jim Thoren, 7311-75th Cir. N., Mpls., MN 55428. PH: 612-425-1190.
Jan 19 MN, Brooklyn Center. 16th Annual Dolls In Winter Doll, Bear & Miniature Sale. Earle Brown Heritage Ctr., 6155 Earle Brown Dr. (I-694 & Hwy. 100). SH: 10am-4pm. Carol's Doll House, 10761 University Ave. NE, Blaine, MN 55434. PH: 612-755-7475.
Jan 25-26 MN, Mankato. 19th Annual Farm & Antique Toy Show. Civic Ctr. SH: Sat. 12noon-5pm, Sun. 9am-3pm. Dick or Marion Sonnek, 24501 470th Ave., Mapleton, MN 56065. PH: 507-524-3275.
Feb 23 MN, St. Paul. Midwest McDonald's Collectors Show. Bandana Square Mall, 1021 Bandana Blvd. E. SH: 12noon-5pm, T: 50. Marilyn Walther, 7013 Bristol Blvd., Edina, MN 55435. PH: 612-929-2951.
Mar 23 MN, Minneapolis. Northwest Toy Show. Northwest Inn Hotel, Hwy. 694 & Hwy. 81 N. SH: 9:30am-2:30pm, T: 130. Jim Thoren, 7311-75th Circle N., Mpls., MN 55428. PH: 612-425-1190.
Mar 29 MN, Duluth. "Collector's Expo". Grandma's Sports Garden, 425 Lake Ave. S. SH: 10am-5pm, T: 32-8'. Timothy Broman, 1600 Miller Trunk Hwy. #112, Duluth, MN 55811. PH: 218-726-1360.
Apr 20 MN, Minneapolis. Bloomington Toy Show. K of C Hall, 1114 W. 79th St. SH: 9:30am-2:30pm, T: 95. Jim Thoren, 7311-75th Circle N., Mpls., MN 55428. PH: 612-425-1190.
May 4 MN, St. Paul. 21st Annual Northland Antique Toy Show. State Fairgrounds, Progress Ctr. SH: 10am-4pm. Carole Mayers, PH: 612-644-1866 or Ray Lacktorin, PH: 612-429-9401.
Jul 12 MN, New Richland. Toy Show. High School, 306 Ash Ave. S. SH: 9am-4pm, T: 100. Rich Lorenz, 14495 State Highway 30, New Richland, MN 56072. PH: 507-465-8368.

Jul 27 MN, Minneapolis. "Barbie Madness Mega Show". Western Thunderbird Hotel, 2201 E. 78th St. SH: 10am-4pm. PH: 612-854-3411.
Aug 30 MN, Duluth. "Collector's Expo". Grandma's Sports Garden, 425 Lake Ave. S. SH: 10am-5pm, T: 32-8'. Timothy Broman, 1600 Miller Trunk Hwy. #112, Duluth, MN 55811. PH: 218-726-1360.
Sep 21 MN, Minneapolis. Northwest Toy Show. Northwest Inn Hotel, Hwy. 694 & Hwy. 81 N. SH: 9:30am-2:30pm, T: 130. Jim Thoren, 7311-75th Circle N., Mpls., MN 55428. PH: 612-425-1190.
Sep 29 MN, Rochester. 1st Annual Dolls In Fall Doll, Bear & Miniature Sale. Civic Ctr., 30 Civic Ctr. DR SE. SH: 10am-4pm. Carol's Doll House, 10761 University Ave. NE., Blaine, MN 55434. PH: 612-755-7475.
Oct 19 MN, Minneapolis. Bloomington Toy Show. K of C Hall, 1114 W. 79th St. SH: 9:30am-2:30pm, T: 95. Jim Thoren, 7311-75th Circle N., Mpls., MN 55428. PH: 612-425-1190.

MISSOURI

Feb 1-2 MO, St. Louis. Gateway Mid-America Toy Show. Airport North Holiday Inn, I-70 & Lindbergh. T: 300. Roy Baker, RR 1, Box 88, Shipman, IL 62685. PH: 618-836-7787.
May 4 MO, St. Louis. 8th Annual Show. Airport Marriott, I-70 at Lambert International Airport, S. of Exit 236. SH: 9am-3pm, T: 200. Bill McCormick, 752A Woodside Trls., St. Louis, MO 63021. PH: 314-230-3181.

NEBRASKA

Feb 23 NE, Omaha. 7th Annual Toy Collectors' Festival. Millard Social Hall, 10508 S. 144th, Exit 440 on I-80. SH: 9am-3pm. Don Armstrong, PH: 402-289-2195.

NEVADA

Jan 4-5 NV, Henderson-Las Vegas. Dolls, Bears, Supplies & Crafts Show, Henderson Convention Center, 200 Water St. SH: Sat. 10am-5pm, Sun. 11am-4pm. RSM Productions, PO Box 967, Rancho Santa Fe, CA 92067. PH: 619-756-3275.
Feb 1 NV, Las Vegas. Barbie Visits Las Vegas. Santa Fe Hotel & Casino, 4949 N. Rancho Dr. Heavenly Doll Company, PO Box 774, S. Lake Tahoe, CA 96156. PH: 702-588-DOLL or FAX: 702-588-BRBI.
Mar 23 NV, Las Vegas. "Barbie Goes To Las Vegas". Gold Coast, 4000 W. Flamingo Rd. SH: 10am-4pm. Marl, PH: 941-751-6275 or Joe, PH: 213-953-6490.
Jun 7 NV, Reno. Toys, Dolls, Bears & All Collectibles. Sundowner Hotel, 450 N. Arlington. Heavenly Doll Company, PO Box 774, S. Lake Tahoe, CA 96156. PH: 702-588-DOLL or FAX: 702-588-BRBI.
Oct 25-26 NV, Henderson-Las Vegas. Dolls, Bears, Supplies & Crafts Show, Henderson Convention Center, 200 Water St. SH: Sat. 10am-5pm, Sun. 11am-4pm. RSM Productions, PO Box 967, Rancho Santa Fe, CA 92067. PH: 619-756-3275.

NEW HAMPSHIRE

Feb 9 NH, Nashua. 1st All Barbie Show. Sheraton Tara Hotel, Exit 1, Rt. 3. SH: 10am-3pm. Cindy Amburgey, 47 Jackson Ave., Fitchburg, MA 01420. PH: 508-342-8292 or 342-2265.
Mar 9 NH, Nashua. Doll & Bear Show. Marriott Hotel, Exit 8, Rt. 3. SH: 10am-3pm. Cindy Amburgey, 47 Jackson Ave., Fitchburg, MA 01420. PH: 508-342-8292 or 342-2265.

NEW JERSEY

Jan 1 NJ, East Brunswick. New Year's Toys, Cards, Comics & Collectibles Show. Ramada Inn, 195 Route 18 S. at NJ Tpke. Exit 9. SH: 11am-5pm, T: 40-45 6' & 8'. Sallie Natowitz, PO Box 796, Matawan, NJ 07747. PH: 908-583-7915.
Jan 5 NJ, Garfield. Toys, Comics, Cards & Collectibles Show. Boys & Girls Club, 490 Midland Ave. SH: 9:30am-3:30pm, T: 60-8'. Gary Sipos, 41 Cedar St., Garfield, NJ 07026. PH: 201-340-2218 or 478-7662 day of show.
Jan 5 NJ, E. Hanover. Big Toy Event Antique & Collectible Show. Ramada Hotel, 130 Rt. 10 W. SH: 10am-4pm, T: 125. Heritage Promos., Yolanda Stanczyk, PO Box 368 Tannersville, PA 18372. PH: 717-620-2422.
Jan 5 NJ, Wayne. Collectibles Show. Co. #1 Firemens Convention Ctr., 1 Parish Dr. off 23N at Rt. 202. SH: 9am-4pm, T: 110-6'. Phil DeMario, 137 Silleck St., Clifton, NJ 07013. PH: 201-742-2647 or 694-9773 day of show.
Jan 11 NJ, Cherry Hill. Toys, Sports Card & Comic Book Show. Holiday Inn, Rt. 70 & Sayre Ave. SH: 9:30am-3:30pm, T: 40-6'. Premier Promos., Greg Petrino, 93 Grant Dr., Holland, PA 18966. PH: 215-968-8577 or Eric Poppel, PH: 215-579-0731.
Jan 12 NJ, Elmwood Park. Toys, Cards, Comics & Collectibles Show. American Legion Post #147, Legion Pl., 1 blk. W. of Midland Ave., off Market St. SH: 9:30am-3:30pm, T: 50-6'. Gary Sipos, 41 Cedar St., Garfield, NJ 07026. PH: 201-340-2218 or 796-9717 day of show.
Jan 12 NJ, Livingston. Train Show. Travelodge, 550 W. Mt. Pleasant Ave. (Rt. 10). SH: 10am-4pm. Key Promos., Ltd., PO Box 51, Metuchen, NJ 08840. PH: 908-756-2385 or 233-7949.
Jan 19 NJ, Wayne. Toy & Collectible Show, Wayne Manor, Grandball Room. 1515 Rt. 23 S., SH: 8am-3pm, T: 100-7', F: $40., A: $4., under 12 free. Phil DeMario, 137 Silleck St., Clifton, NJ 07013. PH: 201-742-2647.
Jan 19 NJ, Garfield. Toys, Comics, Cards & Collectibles Show. Boys & Girls Club, 490 Midland Ave. SH: 9:30am-3:30pm, T: 60-8'. Gary Sipos, 41 Cedar St., Garfield, NJ 07026. PH: 201-340-2218 or 478-7662 day of show.
Jan 19 NJ, E. Hanover. Greater NE Doll & Teddy Bear Show. Ramada Hotel, 130 Route 10 W. SH: 10am-4pm, T: 150. Heritage Promos., Yolanda Stanczyk, PO Box 368, Tannersville, PA 18372. PH: 717-620-2422.
Jan 23 NJ, Elmwood Park. Toys, Cards, Comics & Collectibles Show. American Legion Post #147, Legion Pl., 1 blk. W. of Midland Ave., on Market St. SH: 5pm-10pm, T: 40-6'. Gary Sipos, 41 Cedar St., Garfield, NJ 07026. PH: 201-340-2218 or 796-9717 day of show.
Jan 26 NJ, Wayne. The Great Bergen Passaic Toy & Train Show. P.A.L. Hall. SH: 9am-2pm. PH: 718-229-6282.
Jan 26 NJ, Clifton. Toys, Comics, Cards & Collectibles Show. Boys & Girls Club, 802 Clifton Ave. SH: 9:30am-3:30pm, T: 50-8'. Gary Sipos, 41 Cedar St., Garfield, NJ 07026. PH: 201-340-2218 or 779-9332 day of show.
Feb 2 NJ, Newark. World of Mini Mania Miniature & Miniature Doll Show-Winter Expo. Holiday Inn North, 160 Holiday Plaza, Exit Frontage Rd., Exit 14 from NJ Tpke. SH: 10am-4:30pm, no strollers. Victorian Vintage, PO Box 761, Clark, NJ 07066 (SASE), PH: 908-382-2135. FAX: 908-382-1910.
Feb 2 NJ, Livingston. Doll Show. Travelodge, 550 W. Mt. Pleasant Ave. (Rt. 10). SH: 10am-4pm. Key Promos., Ltd., PO Box 51, Metuchen, NJ 08840. PH: 908-756-2385 or 233-7949.
Feb 2 NJ, Garfield. Toys, Comics, Cards & Collectibles Show. Boys & Girls Club, 490 Midland Ave. SH: 9:30am-3:30pm, T: 60-8'. Gary Sipos, 41 Cedar St., Garfield, NJ 07026. PH: 201-340-2218 or 478-7662 day of show.
Feb 2 NJ, Wayne. Collectibles Show. Co. #1 Firemens Convention Ctr., 1 Parish Dr. off 23N at Rt. 202. SH: 9am-4pm, T: 110-6'. Phil DeMario, 137 Silleck St., Clifton, NJ 07013. PH: 201-742-2647 or 694-9773 day of show.
Feb 9 NJ, Elmwood Park. Toys, Cards, Comics & Collectibles Show. American Legion Post #147, Legion Pl., 1 blk. W. of Midland Ave., off Market St. SH: 9:30am-3:30pm, T: 50-6'. Gary Sipos, 41 Cedar St., Garfield, NJ 07026. PH: 201-340-2218 or 796-9717 day of show.
Feb 16 NJ, Garfield. Toys, Comics, Cards & Collectibles Show. Boys & Girls Club, 490 Midland Ave. SH: 9:30am-3:30pm, T: 60-8'. Gary Sipos, 41 Cedar St., Garfield, NJ 07026. PH: 201-340-2218 or 478-7662 day of show.
Feb 20 NJ, Elmwood Park. Toys, Cards, Comics & Collectibles Show. American Legion Post #147, Legion Pl., 1 blk. W. of Midland Ave., on Market St. SH: 5pm-10pm, T: 40-6'. Gary Sipos, 41 Cedar St., Garfield, NJ 07026. PH: 201-340-2218 or 796-9717 day of show.
Feb 22-23 NJ, Pennsauken. Greenberg's Great Train, Dollhouse & Toy Show. South Jersey Expo Ctr., 2323 Route 73. SH: Sat. 11am-5pm, Sun. 11am-4pm. Greenberg Shows, 7566 Main St., Sykesville, MD 21784. PH: 410-795-7447.
Feb 23 NJ, Livingston. Toy Show. Travelodge, 550 W. Mt. Pleasant Ave. (Rt. 10). SH: 10am-4pm. Key Promos., Ltd., PO Box 51, Metuchen, NJ 08840. PH: 908-756-2385 or 233-7949.
Feb 23 NJ, Clifton. Toys, Comics, Cards & Collectibles Show. Boys & Girls Club, 802 Clifton Ave. SH: 9:30am-3:30pm, T: 50-8'. Gary Sipos, 41 Cedar St., Garfield, NJ 07026. PH: 201-340-2218 or 779-9332 day of show.
Mar 1-2 NJ, Edison. Greenberg's Great Train, Dollhouse & Toy Show. NJ Convention & Expo Ctr., 97 Sunfield Ave. SH: Sat. 11am-5pm, Sun. 11am-4pm. Greenberg Shows, 7566 Main St., Sykesville, MD 21784. PH: 410-795-7447.
Mar 2 NJ, Middletown Twsp. Central Jersey Toy & Model Car Show. VFW #2179, Rt. 36 E., GSP Exit 117 to Rt. 36 E. (Sandy Hook) R. onto Vet. Ln. SH: 9:30am-3pm, T: 150. R. Fusella, PO Box 351, Leonardo, NJ 07737. PH: 908-872-7191.
Mar 2 NJ, Elmwood Park. Toys, Cards, Comics & Collectibles Show. American Legion Post #147, Legion Pl., 1 blk. W. of Midland Ave., off Market St. SH: 9:30am-3:30pm, T: 50-6'. Gary Sipos, 41 Cedar St., Garfield, NJ 07026. PH: 201-340-2218 or 796-9717 day of show.
Mar 2 NJ, Wayne. Collectibles Show. Co. #1 Firemens Convention Ctr., 1 Parish Dr. off 23N at Rt. 202. SH: 9am-4pm, T: 110-6'. Phil DeMario, 137 Silleck St., Clifton, NJ 07013. PH: 201-742-2647 or 694-9773 day of show.
Mar 9 NJ, Garfield. Toys, Comics, Cards & Collectibles Show. Boys & Girls Club, 490 Midland Ave. SH: 9:30am-3:30pm, T: 60-8'. Gary Sipos, 41 Cedar St., Garfield, NJ 07026. PH: 201-340-2218 or 478-7662 day of show.
Mar 16 NJ, North Bergen. Toy Soldier, Military Toys & Action Figure Show. Schuetzen Park, 32nd St. & Kennedy Blvd. SH: 9am-4pm, T: 100. Ed Gries, PO Box 572, Hackensack, NJ 07602. PH: 201-342-6475.
Mar 16 NJ, Hasbrouck Heights. Collectible Toy Show. VFW Hall, (behind IHOP on 17 S.). SH: 9:30am-3:30pm, T: 65. Phil, PH: 201-772-5466.
Mar 16 NJ, Garfield. Toys, Comics, Cards & Collectibles Show. Boys & Girls Club, 490 Midland Ave. SH: 9:30am-3:30pm, T: 60-8'. Gary Sipos, 41 Cedar St., Garfield, NJ 07026. PH: 201-340-2218 or 478-7662 day of show.
Mar 20 NJ, Elmwood Park. Toys, Cards, Comics & Collectibles Show. American Legion Post #147, Legion Pl., 1 blk. W. of Midland Ave., on Market St. SH: 5pm-10pm, T: 40-6'. Gary Sipos, 41 Cedar St., Garfield, NJ 07026. PH: 201-340-2218 or 796-9717 day of show.
Mar 22-23 NJ, Atlantic City. "Atlantique City" Spring Festival-Indoor Antique & Collectibles Show. Convention Center, Boardwalk at Florida Ave. SH: Sat. 10am-9pm, Sun. 10am-5pm, 1200 dealers. Brimfield Associates Inc., PO Box 1800, Ocean City, NJ 08226. PH: 800-526-2724.
Mar 22-23 NJ, Hackensack. Greenberg's Great Train, Dollhouse & Toy Show. Fairleigh-Dickinson Univ., 100 University Plaza Dr. SH: Sat. 11am-5pm, Sun. 11am-4pm. Greenberg Shows, 7566 Main St., Sykesville, MD 21784. PH: 410-795-7447.
Mar 23 NJ, Clifton. Toys, Comics, Cards & Collectibles Show. Boys & Girls Club, 802 Clifton Ave. SH: 9:30am-3:30pm, T: 50-8'. Gary Sipos, 41 Cedar St., Garfield, NJ 07026. PH: 201-340-2218 or 779-9332 day of show.
Apr 6 NJ, Garfield. Toys, Comics, Cards & Collectibles Show. Boys & Girls Club, 490 Midland Ave. SH: 9:30am-3:30pm, T: 60-8'. Gary Sipos, 41 Cedar St., Garfield, NJ 07026. PH: 201-340-2218 or 478-7662 day of show.
Apr 6 NJ, Wayne. Collectibles Show. Co. #1 Firemens Convention Ctr., 1 Parish Dr. off 23N at Rt. 202. SH: 9am-4pm, T: 110-6'. Phil DeMario, 137 Silleck St., Clifton, NJ 07013. PH: 201-742-2647 or 694-9773 day of show.
Apr 13 NJ, Wayne. The Great Bergen Passaic Toy & Train Show. P.A.L. Hall. SH: 9am-2pm. PH: 718-229-6282.
Apr 13 NJ, Elmwood Park. Toys, Cards, Comics & Collectibles Show. American Legion Post #147, Legion Pl., 1 blk. W. of Midland Ave., off Market St. SH: 9:30am-3:30pm, T: 50-6'. Gary Sipos, 41 Cedar St., Garfield, NJ 07026. PH: 201-340-2218 or 796-9717 day of show.
Apr 20 NJ, Garfield. Toys, Comics, Cards & Collectibles Show. Boys & Girls Club, 490 Midland Ave. SH: 9:30am-3:30pm, T: 60-8'. Gary Sipos, 41 Cedar St., Garfield, NJ 07026. PH: 201-340-2218 or 478-7662 day of show.
Apr 24 NJ, Elmwood Park. Toys, Cards, Comics & Collectibles Show. American Legion Post #147, Legion Pl., 1 blk. W. of Midland Ave., on Market St. SH: 5pm-10pm, T: 40-6'. Gary Sipos, 41 Cedar St., Garfield, NJ 07026. PH: 201-340-2218 or 796-9717 day of show.
Apr 26 NJ, Pole Tavern. Spring Doll Show. Elmer Grange Hall, Daretown Rd. SH: 10am-4pm, T: 40. Judy Penza, 375 Route 206, Hammonton, NJ 08037. PH: 609-561-9122.
Apr 27 NJ, Clifton. Toys, Comics, Cards & Collectibles Show. Boys & Girls Club, 802 Clifton Ave. SH: 9:30am-3:30pm, T: 50-8'. Gary Sipos, 41 Cedar St., Garfield, NJ 07026. PH: 201-340-2218 or 779-9332 day of show.
May 4 NJ, Middletown Twsp. Central Jersey Toy & Model Car Show. VFW #2179, Rt. 36 E., GSP Exit 117 to Rt. 36 E. (Sandy Hook) R. onto Vet. Ln. SH: 9:30am-3pm, T: 150. R. Fusella, PO Box 351, Leonardo, NJ 07737. PH: 908-872-7191.
May 4 NJ, Garfield. Toys, Comics, Cards & Collectibles Show. Boys & Girls Club, 490 Midland Ave. SH: 9:30am-3:30pm, T: 60-8'. Gary Sipos, 41 Cedar St., Garfield, NJ 07026. PH: 201-340-2218 or 478-7662 day of show.
May 4 NJ, Wayne. Collectibles Show. Co. #1 Firemens Convention Ctr., 1 Parish Dr. off 23N at Rt. 202. SH: 9am-4pm, T: 110-6'. Phil DeMario, 137 Silleck St., Clifton, NJ 07013. PH: 201-742-2647 or 694-9773 day of show.
May 11 NJ, Elmwood Park. Toys, Cards, Comics & Collectibles Show. American Legion Post #147, Legion Pl., 1 blk. W. of Midland Ave., off Market St. SH: 9:30am-3:30pm, T: 50-6'. Gary Sipos, 41 Cedar St., Garfield, NJ 07026. PH: 201-340-2218 or 796-9717 day of show.
May 18 NJ, Garfield. Toys, Comics, Cards & Collectibles Show. Boys & Girls Club, 490 Midland Ave. SH: 9:30am-3:30pm, T: 60-8'. Gary Sipos, 41 Cedar St., Garfield, NJ 07026. PH: 201-340-2218 or 478-7662 day of show.
May 22 NJ, Elmwood Park. Toys, Cards, Comics & Collectibles Show. American Legion Post #147, Legion Pl., 1 blk. W. of Midland Ave., on Market St. SH: 5pm-10pm, T: 40-6'. Gary Sipos, 41 Cedar St., Garfield, NJ 07026. PH: 201-340-2218 or 796-9717 day of show.
May 25 NJ, Clifton. Toys, Comics, Cards & Collectibles Show. Boys & Girls Club, 802 Clifton Ave. SH: 9:30am-3:30pm, T: 50-8'. Gary Sipos, 41 Cedar St., Garfield, NJ 07026. PH: 201-340-2218 or 779-9332 day of show.
Jun 1 NJ, Garfield. Toys, Comics, Cards & Collectibles Show. Boys & Girls Club, 490 Midland Ave. SH: 9:30am-3:30pm, T: 60-8'. Gary Sipos, 41 Cedar St., Garfield, NJ 07026. PH: 201-340-2218 or 478-7662 day of show.
Jun 1 NJ, Wayne. Collectibles Show. Co. #1 Firemens Convention Ctr., 1 Parish Dr. off 23N at Rt. 202. SH: 9am-4pm, T: 110-6'. Phil DeMario, 137 Silleck St., Clifton, NJ 07013. PH: 201-742-2647 or 694-9773 day of show.
Jun 8 NJ, Elmwood Park. Toys, Cards, Comics & Collectibles Show. American Legion Post #147, Legion Pl., 1 blk. W. of Midland Ave., off Market St. SH: 9:30am-3:30pm, T: 50-6'. Gary Sipos, 41 Cedar St., Garfield, NJ 07026. PH: 201-340-2218 or 796-9717 day of show.
Jun 15 NJ, Hasbrouck Heights. Collectible Toy Show. VFW Hall, (behind IHOP on 17 S.). SH: 9:30am-3:30pm, T: 65. Phil, PH: 201-772-5466.
Jun 15 NJ, Garfield. Toys, Comics, Cards & Collectibles Show. Boys & Girls Club, 490 Midland Ave. SH: 9:30am-3:30pm, T: 60-8'. Gary Sipos, 41 Cedar St., Garfield, NJ 07026. PH: 201-340-2218 or 478-7662 day of show.
Jun 19 NJ, Elmwood Park. Toys, Cards, Comics & Collectibles Show. American Legion Post #147, Legion Pl., 1 blk. W. of Midland Ave., on Market St. SH: 5pm-10pm, T: 40-6'. Gary Sipos, 41 Cedar St., Garfield, NJ 07026. PH: 201-340-2218 or 796-9717 day of show.
Jun 22 NJ, Clifton. Toys, Comics, Cards & Collectibles Show. Boys & Girls Club, 802 Clifton Ave. SH: 9:30am-3:30pm, T: 50-8'. Gary Sipos, 41 Cedar St., Garfield, NJ 07026. PH: 201-340-2218 or 779-9332 day of show.
Jun 29 NJ, Garfield. Toys, Comics, Cards & Collectibles Show. Boys & Girls Club, 490 Midland Ave. SH: 9:30am-3:30pm, T: 60-8'. Gary Sipos, 41 Cedar St., Garfield, NJ 07026. PH: 201-340-2218 or 478-7662 day of show.
Jul 6 NJ, Wayne. Collectibles Show. Co. #1 Firemens Convention Ctr., 1 Parish Dr. off 23N at Rt. 202. SH: 9am-4pm, T: 110-6'. Phil DeMario, 137 Silleck St., Clifton, NJ 07013. PH: 201-742-2647 or 694-9773 day of show.
Jul 13 NJ, Elmwood Park. Toys, Cards, Comics & Collectibles Show. American Legion Post #147, Legion Pl., 1 blk. W. of Midland Ave., off Market St. SH: 9:30am-3:30pm, T: 50-6'. Gary Sipos, 41 Cedar St., Garfield, NJ 07026. PH: 201-340-2218 or 796-9717 day of show.
Jul 13 NJ, Secaucus. "Barbie Madness Mega Show". Meadowlands Hilton, 2 Harmon Plaza. SH: 10am-4pm. PH: 201-348-6900.
Jul 20 NJ, Garfield. Toys, Comics, Cards & Collectibles Show. Boys & Girls Club, 490 Midland Ave. SH: 9:30am-3:30pm, T: 60-8'. Gary Sipos, 41 Cedar St., Garfield, NJ 07026. PH: 201-340-2218 or 478-7662 day of show.
Jul 24 NJ, Elmwood Park. Toys, Cards, Comics & Collectibles Show. American Legion Post #147, Legion Pl., 1 blk. W. of Midland Ave., on Market St. SH: 5pm-10pm, T: 40-6'. Gary Sipos, 41 Cedar St., Garfield, NJ 07026. PH: 201-340-2218 or 796-9717 day of show.
Jul 27 NJ, Middletown Twsp. Central Jersey Toy & Model Car Show. VFW #2179, Rt. 36 E., GSP Exit 117 to Rt. 36 E. (Sandy Hook) R. onto Vet. Ln. SH: 9:30am-3pm, T: 150. R. Fusella, PO Box 351, Leonardo, NJ 07737. PH: 908-872-7191.
Jul 27 NJ, Clifton. Toys, Comics, Cards & Collectibles Show. Boys & Girls Club, 802 Clifton Ave. SH: 9:30am-3:30pm, T: 50-8'. Gary Sipos, 41 Cedar St., Garfield, NJ 07026. PH: 201-340-2218 or 779-9332 day of show.
Aug 3 NJ, Garfield. Toys, Comics, Cards & Collectibles Show. Boys & Girls Club, 490 Midland Ave. SH: 9:30am-3:30pm, T: 60-8'. Gary Sipos, 41 Cedar St., Garfield, NJ 07026. PH: 201-340-2218 or 478-7662 day of show.

Aug 3 NJ, Wayne. Collectibles Show. Co. #1 Firemens Convention Ctr., 1 Parish Dr. off 23N at Rt. 202. SH: 9am-4pm, T: 110-6'. Phil DeMario, 137 Silleck St., Clifton, NJ 07013. PH: 201-742-2647 or 694-9773 day of show.

Aug 10 NJ, Elmwood Park. Toys, Cards, Comics & Collectibles Show. American Legion Post #147, Legion Pl., 1 blk. W. of Midland Ave., off Market St. SH: 9:30am-3:30pm, T: 50-6'. Gary Sipos, 41 Cedar St., Garfield, NJ 07026. PH: 201-340-2218 or 796-9717 day of show.

Aug 17 NJ, Garfield. Toys, Comics, Cards & Collectibles Show. Boys & Girls Club, 490 Midland Ave. SH: 9:30am-3:30pm, T: 60-8'. Gary Sipos, 41 Cedar St., Garfield, NJ 07026. PH: 201-340-2218 or 478-7662 day of show.

Aug 21 NJ, Elmwood Park. Toys, Cards, Comics & Collectibles Show. American Legion Post #147, Legion Pl., 1 blk. W. of Midland Ave., on Market St. SH: 5pm-10pm, T: 40-6'. Gary Sipos, 41 Cedar St., Garfield, NJ 07026. PH: 201-340-2218 or 796-9717 day of show.

Aug 24 NJ, Clifton. Toys, Comics, Cards & Collectibles Show. Boys & Girls Club, 802 Clifton Ave. SH: 9:30am-3:30pm, T: 50-8'. Gary Sipos, 41 Cedar St., Garfield, NJ 07026. PH: 201-340-2218 or 779-9332 day of show.

Sep 7 NJ, Wayne. The Great Bergen Passaic Toy & Train Show. P.A.L. Hall. SH: 9am-2pm. PH: 718-229-6282.

Sep 7 NJ, Garfield. Toys, Comics, Cards & Collectibles Show. Boys & Girls Club, 490 Midland Ave. SH: 9:30am-3:30pm, T: 60-8'. Gary Sipos, 41 Cedar St., Garfield, NJ 07026. PH: 201-340-2218 or 478-7662 day of show.

Sep 7 NJ, Wayne. Collectibles Show. Co. #1 Firemens Convention Ctr., 1 Parish Dr. off 23N at Rt. 202. SH: 9am-4pm, T: 110-6'. Phil DeMario, 137 Silleck St., Clifton, NJ 07013. PH: 201-742-2647 or 694-9773 day of show.

Sep 14 NJ, Elmwood Park. Toys, Cards, Comics & Collectibles Show. American Legion Post #147, Legion Pl., 1 blk. W. of Midland Ave., off Market St. SH: 9:30am-3:30pm, T: 50-6'. Gary Sipos, 41 Cedar St., Garfield, NJ 07026. PH: 201-340-2218 or 796-9717 day of show.

Sep 14 NJ, Newark. World of Mini Mania Miniature & Miniature Doll Show-Winter Frolic. Holiday Inn North, Exit at Frontage Rd. off of Rt. 1. SH: 10am-4:30pm, no strollers. Victorian Vintage, PO Box 761, Clark, NJ 07066 (SASE), PH: 908-382-2135. FAX: 908-382-1910.

Sep 21 NJ, Hasbrouck Heights. Collectible Toy Show. VFW Hall, (behind IHOP on 17 S.). SH: 9:30am-3:30pm, T: 65. Phil, PH: 201-772-5466.

Sep 21 NJ, Garfield. Toys, Comics, Cards & Collectibles Show. Boys & Girls Club, 490 Midland Ave. SH: 9:30am-3:30pm, T: 60-8'. Gary Sipos, 41 Cedar St., Garfield, NJ 07026. PH: 201-340-2218 or 478-7662 day of show.

Sep 25 NJ, Elmwood Park. Toys, Cards, Comics & Collectibles Show. American Legion Post #147, Legion Pl., 1 blk. W. of Midland Ave., on Market St. SH: 5pm-10pm, T: 40-6'. Gary Sipos, 41 Cedar St., Garfield, NJ 07026. PH: 201-340-2218 or 796-9717 day of show.

Sep 28 NJ, Clifton. Toys, Comics, Cards & Collectibles Show. Boys & Girls Club, 802 Clifton Ave. SH: 9:30am-3:30pm, T: 50-8'. Gary Sipos, 41 Cedar St., Garfield, NJ 07026. PH: 201-340-2218 or 779-9332 day of show.

Oct 5 NJ, Middletown Twsp. Central Jersey Toy & Model Car Show. VFW #2179, Rt. 36 E., GSP Exit 117 to Rt. 36 E. (Sandy Hook) R. onto Vet. Ln. SH: 9:30am-3pm, T: 150. R. Fusella, PO Box 351, Leonardo, NJ 07737. PH: 908-872-7191.

Oct 5 NJ, Garfield. Toys, Comics, Cards & Collectibles Show. Boys & Girls Club, 490 Midland Ave. SH: 9:30am-3:30pm, T: 60-8'. Gary Sipos, 41 Cedar St., Garfield, NJ 07026. PH: 201-340-2218 or 478-7662 day of show.

Oct 5 NJ, Wayne. Collectibles Show. Co. #1 Firemens Convention Ctr., 1 Parish Dr. off 23N at Rt. 202. SH: 9am-4pm, T: 110-6'. Phil DeMario, 137 Silleck St., Clifton, NJ 07013. PH: 201-742-2647 or 694-9773 day of show.

Oct 12 NJ, Wayne. The Great Bergen Passaic Toy & Train Show. P.A.L. Hall. SH: 9am-2pm. PH: 718-229-6282.

Oct 12 NJ, Elmwood Park. Toys, Cards, Comics & Collectibles Show. American Legion Post #147, Legion Pl., 1 blk. W. of Midland Ave., off Market St. SH: 9:30am-3:30pm, T: 50-6'. Gary Sipos, 41 Cedar St., Garfield, NJ 07026. PH: 201-340-2218 or 796-9717 day of show.

Oct 19 NJ, Garfield. Toys, Comics, Cards & Collectibles Show. Boys & Girls Club, 490 Midland Ave. SH: 9:30am-3:30pm, T: 60-8'. Gary Sipos, 41 Cedar St., Garfield, NJ 07026. PH: 201-340-2218 or 478-7662 day of show.

Oct 23 NJ, Elmwood Park. Toys, Cards, Comics & Collectibles Show. American Legion Post #147, Legion Pl., 1 blk. W. of Midland Ave., on Market St. SH: 5pm-10pm, T: 40-6'. Gary Sipos, 41 Cedar St., Garfield, NJ 07026. PH: 201-340-2218 or 796-9717 day of show.

Oct 26 NJ, Clifton. Toys, Comics, Cards & Collectibles Show. Boys & Girls Club, 802 Clifton Ave. SH: 9:30am-3:30pm, T: 50-8'. Gary Sipos, 41 Cedar St., Garfield, NJ 07026. PH: 201-340-2218 or 779-9332 day of show.

Nov 2 NJ, Wayne. Collectibles Show. Co. #1 Firemens Convention Ctr., 1 Parish Dr. off 23N at Rt. 202. SH: 9am-4pm, T: 110-6'. Phil DeMario, 137 Silleck St., Clifton, NJ 07013. PH: 201-742-2647 or 694-9773 day of show.

Nov 9 NJ, Elmwood Park. Toys, Cards, Comics & Collectibles Show. American Legion Post #147, Legion Pl., 1 blk. W. of Midland Ave., off Market St. SH: 9:30am-3:30pm, T: 50-6'. Gary Sipos, 41 Cedar St., Garfield, NJ 07026. PH: 201-340-2218 or 796-9717 day of show.

Nov 16 NJ, Garfield. Toys, Comics, Cards & Collectibles Show. Boys & Girls Club, 490 Midland Ave. SH: 9:30am-3:30pm, T: 60-8'. Gary Sipos, 41 Cedar St., Garfield, NJ 07026. PH: 201-340-2218 or 478-7662 day of show.

Nov 20 NJ, Elmwood Park. Toys, Cards, Comics & Collectibles Show. American Legion Post #147, Legion Pl., 1 blk. W. of Midland Ave., on Market St. SH: 5pm-10pm, T: 40-6'. Gary Sipos, 41 Cedar St., Garfield, NJ 07026. PH: 201-340-2218 or 796-9717 day of show.

Nov 23 NJ, Clifton. Toys, Comics, Cards & Collectibles Show. Boys & Girls Club, 802 Clifton Ave. SH: 9:30am-3:30pm, T: 50-8'. Gary Sipos, 41 Cedar St., Garfield, NJ 07026. PH: 201-340-2218 or 779-9332 day of show.

Nov 30 NJ, Middletown Twsp. Central Jersey Toy & Model Car Show. VFW #2179, Rt. 36 E., GSP Exit 117 to Rt. 36 E. (Sandy Hook) R. onto Vet. Ln. SH: 9:30am-3pm, T: 150. R. Fusella, PO Box 351, Leonardo, NJ 07737. PH: 908-872-7191.

Nov 30 NJ, Garfield. Toys, Comics, Cards & Collectibles Show. Boys & Girls Club, 490 Midland Ave. SH: 9:30am-3:30pm, T: 60-8'. Gary Sipos, 41 Cedar St., Garfield, NJ 07026. PH: 201-340-2218 or 478-7662 day of show.

Dec 7 NJ, Garfield. Toys, Comics, Cards & Collectibles Show. Boys & Girls Club, 490 Midland Ave. SH: 9:30am-3:30pm, T: 60-8'. Gary Sipos, 41 Cedar St., Garfield, NJ 07026. PH: 201-340-2218 or 478-7662 day of show.

Dec 7 NJ, Wayne. Collectibles Show. Co. #1 Firemens Convention Ctr., 1 Parish Dr. off 23N at Rt. 202. SH: 9am-4pm, T: 110-6'. Phil DeMario, 137 Silleck St., Clifton, NJ 07013. PH: 201-742-2647 or 694-9773 day of show.

Dec 14 NJ, Wayne. The Great Bergen Passaic Toy & Train Show. P.A.L. Hall. SH: 9am-2pm. PH: 718-229-6282.

Dec 14 NJ, Elmwood Park. Toys, Cards, Comics & Collectibles Show. American Legion Post #147, Legion Pl., 1 blk. W. of Midland Ave., off Market St. SH: 9:30am-3:30pm, T: 50-6'. Gary Sipos, 41 Cedar St., Garfield, NJ 07026. PH: 201-340-2218 or 796-9717 day of show.

Dec 18 NJ, Elmwood Park. Toys, Cards, Comics & Collectibles Show. American Legion Post #147, Legion Pl., 1 blk. W. of Midland Ave., on Market St. SH: 5pm-10pm, T: 40-6'. Gary Sipos, 41 Cedar St., Garfield, NJ 07026. PH: 201-340-2218 or 796-9717 day of show.

Dec 21 NJ, Garfield. Toys, Comics, Cards & Collectibles Show. Boys & Girls Club, 490 Midland Ave. SH: 9:30am-3:30pm, T: 60-8'. Gary Sipos, 41 Cedar St., Garfield, NJ 07026. PH: 201-340-2218 or 478-7662 day of show.

Dec 28 NJ, Clifton. Toys, Comics, Cards & Collectibles Show. Boys & Girls Club, 802 Clifton Ave. SH: 9:30am-3:30pm, T: 50-8'. Gary Sipos, 41 Cedar St., Garfield, NJ 07026. PH: 201-340-2218 or 779-9332 day of show.

NEW YORK

Jan 3 NY, Brooklyn. Toys, Cards, Comics & Collectibles Show. St. Dominics Church, 20th Ave. & Bayridge Pky. (75th St.). SH: 6pm-10pm, T: 50-9'. Sunrise Prods., Scotty O'Donnell, PO Box 340625, Ryder Station, Brooklyn, NY 11234. PH: 718-251-2075 or 241-6477.

Jan 4 NY, Sayville-LI. Toy, Comic Book, Sports Card & Collectibles Show. Attias Indoor Flea Market, 5750 Sunrise Hwy. (corner of Sunrise & Broadway). SH: 10am-4pm, T: 42. Paul, PH: 516-289-7398.

Jan 5 NY, Hudson. Philmont Model Railroad Club Toy & Train Swap Meet. Moose Lodge, Rt. 9 & Healy Blvd. SH: 9am-2pm. Rick Washburn, 12 Hudson St., Hudson, NY 12534. PH: 518-828-6508.

Jan 5 NY, Brooklyn. Toys, Cards, Comics & Collectibles Show. Temple Hillel, 2164 Ralph Ave. (at Ave. L). SH: 10am-5pm, T: 50. Sunrise Prods., Scotty O'Donnell, PO Box 340625, Ryder Station, Brooklyn, NY 11234. PH: 718-251-2075 or 241-6477.

Jan 9 NY, Deer Park. Monthly Toy, Comic Book & All Sports Card Show. Sons of Italy-Constantino Brumidi Lodge, 2075 Deer Park Ave. SH: 5:30pm-10pm, T: 40. Paul, PH: 516-289-7398.

Jan 12 NY, Elmont-LI. Model Train, Toy & Doll Show. St. Vincent DePaul School Auditorium, 1510 DePaul St. SH: 10am-3pm. Frank Deorio, PH: 516-352-2127.

Jan 12 NY, Yonkers. Nostalgia Int'l Toy & Train Show. Yonkers Raceway. SH: 9am-3pm. PH: 518-392-2660.

Jan 17 NY, Brooklyn. Toys, Cards, Comics & Collectibles Show. St. Dominics Church, 20th Ave. & Bayridge Pky. (75th St.). SH: 6pm-10pm, T: 50-9'. Sunrise Prods., Scotty O'Donnell, PO Box 340625, Ryder Station, Brooklyn, NY 11234. PH: 718-251-2075 or 241-6477.

Jan 18 NY, Plainview. 1st Annual Convention For Charity "Off the Rack, Out of Con-Trol" Show. Holiday Inn, 215 Sunnyside Blvd., Exit 48 on the LIE on the north side. SH: 10am-6pm, T: 30. Barry or Eric, PH: 516-520-0550 M-F 12noon-8pm or 834-2620 leave a msg.

Jan 19 NY, Brooklyn. Toys, Cards, Comics & Collectibles Show. Temple Hillel, 2164 Ralph Ave. (at Ave. L). SH: 10am-5pm, T: 50. Sunrise Prods., Scotty O'Donnell, PO Box 340625, Ryder Station, Brooklyn, NY 11234. PH: 718-251-2075 or 241-6477.

Jan 19 NY, Bayside-NYC. Toy Soldier Show. Adria Hotel & Conference Ctr., 221*-7 Northern Blvd. SH: 10am-3pm, T: 50. Eric Reinikka, PO Box 170-198, Ozone Park, NY 11417. PH: 718-835-9764.

Jan 19 NY, Rockville Centre. Collectorfest Toy Expo. Holiday Inn, 173 Sunrise Hwy. SH: 10am-4pm, T: 100. Paul, PH: 516-520-0975 or FAX: 516-520-0628.

Jan 26 NY, Freeport-LI. Toy Memories Antique Toy Show. Recreation Ctr., 130 E. Merrick Rd. (1/4 mi. W. of Mdbk. Pky.). SH: 10am-4pm, T: 130. Guy De Marco, PO Box 224, West Hempstead, NY 11552. PH: 516-593-8198.

Jan 31 NY, Brooklyn. Toys, Cards, Comics & Collectibles Show. St. Dominics Church, 20th Ave. & Bayridge Pky. (75th St.). SH: 6pm-10pm, T: 50-9'. Sunrise Prods., Scotty O'Donnell, PO Box 340625, Ryder Station, Brooklyn, NY 11234. PH: 718-251-2075 or 241-6477.

Feb 1 NY, Sayville-LI. Toy, Comic Book, Sports Card & Collectibles Show. Attias Indoor Flea Market, 5750 Sunrise Hwy. (corner of Sunrise & Broadway). SH: 10am-4pm, T: 42. Paul, PH: 516-289-7398.

Feb 2 NY, Brooklyn. Toys, Cards, Comics & Collectibles Show. Temple Hillel, 2164 Ralph Ave. (at Ave. L). SH: 10am-5pm, T: 50. Sunrise Prods., Scotty O'Donnell, PO Box 340625, Ryder Station, Brooklyn, NY 11234. PH: 718-251-2075 or 241-6477.

Feb 7 NY, Brooklyn. Toys, Cards, Comics & Collectibles Show. St. Dominics Church, 20th Ave. & Bayridge Pky. (75th St.). SH: 6pm-10pm, T: 50-9'. Sunrise Prods., Scotty O'Donnell, PO Box 340625, Ryder Station, Brooklyn, NY 11234. PH: 718-251-2075 or 241-6477.

Feb 9 NY, White Plains. The Great Westchester Toy & Train Show. County Center. SH: 9am-3pm. PH: 518-392-2660.

Feb 9 NY, Amityville. Barbie Toy Collectible Show. VFW Hall, 300 Broadway (Rt. 110). SH: 10am-4pm, T: 16. Carol Fabianski, 28 Lake Dr., Copiague, NY 11726. PH: 516-598-4801.

Feb 9 NY, Manhattan. Toy Fair "Barbie Madness". Norman Thomas High School, 33rd St. at Park Ave. SH: 10am-4pm.

Feb 13 NY, Deer Park. Monthly Toy, Comic Book & All Sports Card Show. Sons of Italy-Constantino Brumidi Lodge, 2075 Deer Park Ave. SH: 5:30pm-10pm, T: 40. Paul, PH: 516-289-7398.

Feb 15-16 NY, Hamburg. Western NY Railway Historical Society Train & Toy Show. Erie Cty. Fairgrounds. SH: 10am-5pm. WNYRHS, Joe Kocsis, Box 502, Tonawanda, NY 14151. PH: 716-693-8512.

Feb 16 NY, Franklin Square-LI. RR Lines Train & Toy Show. VFW Hall, 68 Lincoln Rd. SH: 9am-1pm. Rae Romano, PH: 516-775-4801.

Feb 16 NY, Brooklyn. Toys, Cards, Comics & Collectibles Show. Temple Hillel, 2164 Ralph Ave. (at Ave. L). SH: 10am-5pm, T: 50. Sunrise Prods., Scotty O'Donnell, PO Box 340625, Ryder Station, Brooklyn, NY 11234. PH: 718-251-2075 or 241-6477.

Feb 21 NY, Brooklyn. Toys, Cards, Comics & Collectibles Show. St. Dominics Church, 20th Ave. & Bayridge Pky. (75th St.). SH: 6pm-10pm, T: 50-9'. Sunrise Prods., Scotty O'Donnell, PO Box 340625, Ryder Station, Brooklyn, NY 11234. PH: 718-251-2075 or 241-6477.

Mar 1 NY, Albany. Northland Toy Club 22nd Semi-Annual Show. Polish Community Ctr., Washington Ave. Ext. SH: 9:30am-2:30pm, T: 90. Bill Monberger, 84 St. Stevans Ln., Glenville, NY 12306. PH: 518-399-1076.FAX: 415-459-0827.

Mar 2 NY, Hauppauge. Long Island Antique & Collectible Toy Expo. Wind Watch Marriott-Motor Parkway. SH: 10am-3pm, T: 120. Toy Box Productions, Jeff Tuthill, PO Box 356, Shirley, NY 11967. PH: 516-395-1899.

Mar 2 NY, Brooklyn. Toys, Cards, Comics & Collectibles Show. Temple Hillel, 2164 Ralph Ave. (at Ave. L). SH: 10am-5pm, T: 50. Sunrise Prods., Scotty O'Donnell, PO Box 340625, Ryder Station, Brooklyn, NY 11234. PH: 718-251-2075 or 241-6477.

Mar 2 NY, Freeport-LI. Figurama-Toy Figure Show. Recreation Ctr., 130 E. Merrick Rd. (1/4 mi. W. of Mdbk. Pky.). SH: 10am-3pm, T: 90. V. Pugliese, PO Box 309, Albertson, NY 11507. PH: 516-593-8198.

Mar 2 NY, Rockville Centre. Collectorfest Toy Expo. Holiday Inn, 173 Sunrise Hwy. SH: 10am-4pm, T: 100. Paul, PH: 516-520-0975 or FAX: 516-520-0628.

Mar 7 NY, Brooklyn. Toys, Cards, Comics & Collectibles Show. St. Dominics Church, 20th Ave. & Bayridge Pky. (75th St.). SH: 6pm-10pm, T: 50-9'. Sunrise Prods., Scotty O'Donnell, PO Box 340625, Ryder Station, Brooklyn, NY 11234. PH: 718-251-2075 or 241-6477.

Mar 9 NY, Yonkers. Nostalgia Int'l Toy & Train Show. Yonkers Raceway. SH: 9am-3pm. PH: 518-392-2660.

Mar 9 NY, Binghamton. Toy, Doll & Teddy Bear Show. Heritage Country Club & Conference Ctr., 4301 Watson Blvd. (Exit 70-N off Rt. 17). SH: 10am-4pm, T: 75-100. Central NY Promos., Lyn Lake, 35 Hubbard St., Apt. 1, Cortland, NY 13045. PH: 607-753-8580.

Mar 13 NY, Deer Park. Monthly Toy, Comic Book & All Sports Card Show. Sons of Italy-Constantino Brumidi Lodge, 2075 Deer Park Ave. SH: 5:30pm-10pm, T: 40. Paul, PH: 516-289-7398.

Mar 16 NY, Franklin Square-LI. RR Lines Train & Toy Show. VFW Hall, 68 Lincoln Rd. SH: 9am-1pm. Rae Romano, PH: 516-775-4801.

Mar 16 NY, Brooklyn. Toys, Cards, Comics & Collectibles Show. Temple Hillel, 2164 Ralph Ave. (at Ave. L). SH: 10am-5pm, T: 50. Sunrise Prods., Scotty O'Donnell, PO Box 340625, Ryder Station, Brooklyn, NY 11234. PH: 718-251-2075 or 241-6477.

Mar 21 NY, Brooklyn. Toys, Cards, Comics & Collectibles Show. St. Dominics Church, 20th Ave. & Bayridge Pky. (75th St.). SH: 6pm-10pm, T: 50-9'. Sunrise Prods., Scotty O'Donnell, PO Box 340625, Ryder Station, Brooklyn, NY 11234. PH: 718-251-2075 or 241-6477.

Mar 23 NY, Greenlawn-LI. Model Train & Toy Show. Huntington Moose Lodge, 631 Pulaski Rd. SH: 9am-1pm. Carmelo Sancetta, PO Box 1286-M, Bay Shore, NY 11706. PH: 516-666-6855.

Apr 5 NY, Camillus. "Dollies at the Zoo" Doll & Teddy Bear Show. St. Joseph's Church, 5600 W. Genesee St. SH: 10am-4pm. Peter Mylon, 601 Bradford Pky., Syracuse, NY 13224. PH: 315-446-6246.

Apr 6 NY, Elmont-LI. Model Train, Toy & Doll Show. St. Vincent DePaul School Auditorium, 1510 DePaul St. SH: 10am-3pm. Frank Deorio, PH: 516-352-2127.

Apr 6 NY, Brooklyn. Toys, Cards, Comics & Collectibles Show. Temple Hillel, 2164 Ralph Ave. (at Ave. L). SH: 10am-5pm, T: 50. Sunrise Prods., Scotty O'Donnell, PO Box 340625, Ryder Station, Brooklyn, NY 11234. PH: 718-251-2075 or 241-6477.

Apr 6 NY, Rockville Centre. Collectorfest Toy Expo. Holiday Inn, 173 Sunrise Hwy. SH: 10am-4pm, T: 100. Paul, PH: 516-520-0975 or FAX: 516-520-0628.

Apr 11 NY, Brooklyn. Toys, Cards, Comics & Collectibles Show. St. Dominics Church, 20th Ave. & Bayridge Pky. (75th St.). SH: 6pm-10pm, T: 50-9'. Sunrise Prods., Scotty O'Donnell, PO Box 340625, Ryder Station, Brooklyn, NY 11234. PH: 718-251-2075 or 241-6477.

Apr 12 NY, N. Hartford. Heritage Doll Club Show. First United Meth. Church, 105 Genesee St. SH: 10am-4pm, T: 65. Mary Polera, 1236 Hammond Ave., Utica, NY 13501. PH: 315-735-5628.

Apr 13 NY, Amityville. Action Figure & Toy Collectible Show. VFW Hall, 300 Broadway (Rt. 110). SH: 10am-4pm, T: 16. Carol Fabianski, 28 Lake Dr., Copiague, NY 11726. PH: 516-598-4801.

Apr 20 NY, White Plains. The Great Westchester Toy & Train Show. County Center. SH: 9am-3pm. PH: 518-392-2660.

Apr 20 NY, Brooklyn. Toys, Cards, Comics & Collectibles Show. Temple Hillel, 2164 Ralph Ave. (at Ave. L). SH: 10am-5pm, T: 50. Sunrise Prods., Scotty O'Donnell, PO Box 340625, Ryder Station, Brooklyn, NY 11234. PH: 718-251-2075 or 241-6477.

Apr 25 NY, Brooklyn. Toys, Cards, Comics & Collectibles Show. St. Dominics Church, 20th Ave. & Bayridge Pky. (75th St.). SH: 6pm-10pm, T: 50-9'. Sunrise Prods., Scotty O'Donnell, PO Box 340625, Ryder Station, Brooklyn, NY 11234. PH: 718-251-2075 or 241-6477.

Apr 27 NY, Freeport-LI. Toy Memories Antique Toy Show. Recreation Center, 130 E. Merrick Rd. (1/4 mi. W. of Mdbk. Pky.). SH: 10am-4pm, T: 130. Guy DeMarco, PO Box 224, W. Hempstead, NY 11552. PH: 516-593-8198.

May 4 NY, Oneonta. Chenango Valley Doll Club Doll Show. Holiday Inn. Rt. 23. SH: 10am-4pm, T: 25.

May 4 NY, Bayside-NYC. Military Toy Show XIII. Adria Hotel & Conference Ctr., 221*-7 Northern Blvd. SH: 10am-3pm, T: 50. Eric Reinikka, PO Box 170*-98, Ozone Park, NY 11417. PH: 718-835-9764.

May 16 NY, Brooklyn. Toys, Cards, Comics & Collectibles Show. St. Dominics Church, 20th Ave. & Bayridge Pky. (75th St.). SH: 6pm-10pm, T: 50-9'. Sunrise Prods., Scotty O'Donnell, PO Box 340625, Ryder Station, Brooklyn, NY 11234. PH: 718-251-2075 or 241-6477.

May 18 NY, Brooklyn. Toys, Cards, Comics & Collectibles Show. Temple Hillel, 2164 Ralph Ave. (at Ave. L). SH: 10am-5pm, T: 50. Sunrise Prods., Scotty O'Donnell, PO Box 340625, Ryder Station, Brooklyn, NY 11234. PH: 718-251-2075 or 241-6477.

May 18 NY, Amityville. Barbie Toy Collectible Show. VFW Hall, 300 Broadway (Rt. 110). SH: 10am-4pm, T: 16. Carol Fabianski, 28 Lake Dr., Copiague, NY 11726. PH: 516-598-4801.

May 25 NY, Franklin Square-LI. RR Lines Train & Toy Show. VFW Hall, 68 Lincoln Rd. SH: 9am-1pm. Rae Romano, PH: 516-775-4801.

May 25 NY, Brooklyn. Toys, Cards, Comics & Collectibles Show. Temple Hillel, 2164 Ralph Ave. (at Ave. L). SH: 10am-5pm, T: 50. Sunrise Prods., Scotty O'Donnell, PO Box 340625, Ryder Station, Brooklyn, NY 11234. PH: 718-251-2075 or 241-6477.

May 30 NY, Brooklyn. Toys, Cards, Comics & Collectibles Show. St. Dominics Church, 20th Ave. & Bayridge Pky. (75th St.). SH: 6pm-10pm, T: 50-9'. Sunrise Prods., Scotty O'Donnell, PO Box 340625, Ryder Station, Brooklyn, NY 11234. PH: 718-251-2075 or 241-6477.

Jun 1 NY, Brooklyn. Toys, Cards, Comics & Collectibles Show. Temple Hillel, 2164 Ralph Ave. (at Ave. L). SH: 10am-5pm, T: 50. Sunrise Prods., Scotty O'Donnell, PO Box 340625, Ryder Station, Brooklyn, NY 11234. PH: 718-251-2075 or 241-6477.

Jun 8 NY, Brooklyn. Toys, Cards, Comics & Collectibles Show. Temple Hillel, 2164 Ralph Ave. (at Ave. L). SH: 10am-5pm, T: 50. Sunrise Prods., Scotty O'Donnell, PO Box 340625, Ryder Station, Brooklyn, NY 11234. PH: 718-251-2075 or 241-6477.

Jun 22 NY, Brooklyn. Toys, Cards, Comics & Collectibles Show. Temple Hillel, 2164 Ralph Ave. (at Ave. L). SH: 10am-5pm, T: 50. Sunrise Prods., Scotty O'Donnell, PO Box 340625, Ryder Station, Brooklyn, NY 11234. PH: 718-251-2075 or 241-6477.

Jun 29 NY, Franklin Square-LI. RR Lines Train & Toy Show. VFW Hall, 68 Lincoln Rd. SH: 9am-1pm. Rae Romano, PH: 516-775-4801.

Jul 6 NY, Brooklyn. Toys, Cards, Comics & Collectibles Show. Temple Hillel, 2164 Ralph Ave. (at Ave. L). SH: 10am-5pm, T: 50. Sunrise Prods., Scotty O'Donnell, PO Box 340625, Ryder Station, Brooklyn, NY 11234. PH: 718-251-2075 or 241-6477.

Jul 18-20 NY, Rochester. BotCon '97 (Transformers Convention). Riverside Convention Ctr., 123 E. Main St. SH: Fri. 7pm-10pm, Sat. 9am-10pm, Sun. 9am-6pm. Jon Hartman, 602 Seagraves Ave., Kendallville, IN 46755. PH: 219-347-0355.

Jul 20 NY, Brooklyn. Toys, Cards, Comics & Collectibles Show. Temple Hillel, 2164 Ralph Ave. (at Ave. L). SH: 10am-5pm, T: 50. Sunrise Prods., Scotty O'Donnell, PO Box 340625, Ryder Station, Brooklyn, NY 11234. PH: 718-251-2075 or 241-6477.

Jul 27 NY, Brooklyn. Toys, Cards, Comics & Collectibles Show. Temple Hillel, 2164 Ralph Ave. (at Ave. L). SH: 10am-5pm, T: 50. Sunrise Prods., Scotty O'Donnell, PO Box 340625, Ryder Station, Brooklyn, NY 11234. PH: 718-251-2075 or 241-6477.

Aug 3 NY, Greenlawn-LI. Model Train & Toy Show. Huntington Moose Lodge, 631 Pulaski Rd. SH: 9am-1pm. Carmelo Sancetta, PO Box 1286-M, Bay Shore, NY 11706. PH: 516-666-6855.

Aug 10 NY, Elmont-LI. Model Train, Toy & Doll Show. St. Vincent DePaul School Auditorium, 1510 DePaul St. SH: 10am-3pm. Frank Deorio, PH: 516-352-2127.

Aug 10 NY, Brooklyn. Toys, Cards, Comics & Collectibles Show. Temple Hillel, 2164 Ralph Ave. (at Ave. L). SH: 10am-5pm, T: 50. Sunrise Prods., Scotty O'Donnell, PO Box 340625, Ryder Station, Brooklyn, NY 11234. PH: 718-251-2075 or 241-6477.

Aug 10 NY, Amityville. Action Figure & Toy Collectible Show. VFW Hall, 300 Broadway (Rt. 110). SH: 10am-4pm, T: 16. Carol Fabianski, 28 Lake Dr., Copiague, NY 11726. PH: 516-598-4801.

Aug 17 NY, Franklin Square-LI. RR Lines Train & Toy Show. VFW Hall, 68 Lincoln Rd. SH: 9am-1pm. Rae Romano, PH: 516-775-4801.

Aug 24 NY, Brooklyn. Toys, Cards, Comics & Collectibles Show. Temple Hillel, 2164 Ralph Ave. (at Ave. L). SH: 10am-5pm, T: 50. Sunrise Prods., Scotty O'Donnell, PO Box 340625, Ryder Station, Brooklyn, NY 11234. PH: 718-251-2075 or 241-6477.

Aug 24 NY, Amityville. Barbie Toy Collectible Show. VFW Hall, 300 Broadway (Rt. 110). SH: 10am-4pm, T: 16. Carol Fabianski, 28 Lake Dr., Copiague, NY 11726. PH: 516-598-4801.

Aug 31 NY, Brooklyn. Toys, Cards, Comics & Collectibles Show. Temple Hillel, 2164 Ralph Ave. (at Ave. L). SH: 10am-5pm, T: 50. Sunrise Prods., Scotty O'Donnell, PO Box 340625, Ryder Station, Brooklyn, NY 11234. PH: 718-251-2075 or 241-6477.

Sep 7 NY, Brooklyn. Toys, Cards, Comics & Collectibles Show. Temple Hillel, 2164 Ralph Ave. (at Ave. L). SH: 10am-5pm, T: 50. Sunrise Prods., Scotty O'Donnell, PO Box 340625, Ryder Station, Brooklyn, NY 11234. PH: 718-251-2075 or 241-6477.

Sep 7 NY, Freeport-LI. Figurama-Toy Figure Show. Recreation Ctr., 130 E. Merrick Rd. (1
4 mi. W. of Mdbk. Pky.). SH: 10am-3pm, T: 90. V. Pugliese, PO Box 309, Albertson, NY 11507. PH: 516-593-8198.

Sep 21 NY, Franklin Square-LI. RR Lines Train & Toy Show. VFW Hall, 68 Lincoln Rd. SH: 9am-1pm. Rae Romano, PH: 516-775-4801.

Sep 21 NY, Brooklyn. Toys, Cards, Comics & Collectibles Show. Temple Hillel, 2164 Ralph Ave. (at Ave. L). SH: 10am-5pm, T: 50. Sunrise Prods., Scotty O'Donnell, PO Box 340625, Ryder Station, Brooklyn, NY 11234. PH: 718-251-2075 or 241-6477.

Sep 28 NY, Greenlawn-LI. Model Train & Toy Show. Huntington Moose Lodge, 631 Pulaski Rd. SH: 9am-1pm. Carmelo Sancetta, PO Box 1286-M, Bay Shore, NY 11706. PH: 516-666-6855.

Sep 28 NY, Westchester. World of Mania Miniature & Miniature Doll Show-Fall Festival. Holiday Inn Crowne Plaza, White Plains, 66 Hale Ave., Exit 8 from I-287. SH: 10am-4:30pm, no strollers. Victorian Vintage, PO Box 761, Clark, NJ 07066 (SASE), PH: 908-382-2135. FAX: 908-382-1910.

Oct 12 NY, Binghamton. Toy, Doll & Teddy Bear Show. Heritage Country Club & Conference Ctr., 4301 Watson Blvd. (Exit 70-N off Rt. 17). SH: 10am-4pm, T: 75-100. Central NY Promos., Lyn Lake, 35 Hubbard St., Apt. 1, Cortland, NY 13045. PH: 607-753-8580.

Oct 19 NY, Freeport-LI. Toy Memories Antique Toy Show. Recreation Ctr., 130 E. Merrick Rd. (1/4 mi. W. of Mdbk. Pky.). SH: 10am-4pm, T: 130. Guy De Marco, PO Box 224, W. Hempstead, NY 11552. PH: 516-593-8198.FAX: 415-459-0827.

Oct 26 NY, Elmont-LI. Model Train, Toy & Doll Show. St. Vincent DePaul School Auditorium, 1510 DePaul St. SH: 10am-3pm. Frank Deorio, PH: 516-352-2127.

Nov 1 NY, Albany. Northland Toy Club 23rd Semi-Annual Show. Polish Community Ctr., Washington Ave. Ext. SH: 9:30am-2:30pm, T: 90. Bill Monberger, 84 St. Stevans Ln., Glenville, NY 12306. PH: 518-399-1076.FAX: 415-459-0827.

Nov 2 NY, Franklin Square-LI. RR Lines Train & Toy Show. VFW Hall, 68 Lincoln Rd. SH: 9am-1pm. Rae Romano, PH: 516-775-4801.

Nov 16 NY, Amityville. Barbie Toy Collectible Show. VFW Hall, 300 Broadway (Rt. 110). SH: 10am-4pm, T: 16. Carol Fabianski, 28 Lake Dr., Copiague, NY 11726. PH: 516-598-4801.

Nov 23 NY, Greenlawn-LI. Model Train & Toy Show. Huntington Moose Lodge, 631 Pulaski Rd. SH: 9am-1pm. Carmelo Sancetta, PO Box 1286-M, Bay Shore, NY 11706. PH: 516-666-6855.

Dec 7 NY, Elmont-LI. Model Train, Toy & Doll Show. St. Vincent DePaul School Auditorium, 1510 DePaul St. SH: 10am-3pm. Frank Deorio, PH: 516-352-2127.

Dec 14 NY, Amityville. Action Figure & Toy Collectible Show. VFW Hall, 300 Broadway (Rt. 110). SH: 10am-4pm, T: 16. Carol Fabianski, 28 Lake Dr., Copiague, NY 11726. PH: 516-598-4801.

Dec 21 NY, White Plains. The Great Westchester Toy & Train Show. County Center. SH: 9am-3pm. PH: 518-392-2660.

Dec 21 NY, Greenlawn-LI. Model Train & Toy Show. Huntington Moose Lodge, 631 Pulaski Rd. SH: 9am-1pm. Carmelo Sancetta, PO Box 1286-M, Bay Shore, NY 11706. PH: 516-666-6855.

Dec 28 NY, Franklin Square-LI. RR Lines Train & Toy Show. VFW Hall, 68 Lincoln Rd. SH: 9am-1pm. Rae Romano, PH: 516-775-4801.

NORTH CAROLINA

Jan 18-19 NC, Raleigh. 24th Annual North State Toy Collectors Show. State Fairgrounds, just off I-40. SH: Sat. 9am-5pm, Sun. 10am-4:30pm, T: 285. Carolina Hobby Expo, PH: 704-786-8373 after 4pm.

Jan 25 NC, Concord. Toy Car & NASCAR Collectibles Show. Nat'l. Guard Armory, Hwy. 49 & Old Charlotte Rd. SH: 9am-3:30pm, T: 105. Carolina Hobby Expo, PH: 704-786-8373 after 4pm.

Feb 1 NC, Clemmons. 9th Annual Toy Collectors Show. Ramada Inn, I-40, Exit 184. SH: 9am-4pm, T: 132. Carolina Hobby Expo, PH: 704-786-8373 after 4pm.

Feb 2 NC, Charlotte. Festival of Dollhouse Miniatures. Hilton Hotel. Molly Cromwell, 4701 Duncan Dr., Annandale, VA 22003. PH: 703-978-5353.

Feb 8 NC, Fayetteville. 19th Annual Sandhills Toy, Hobby & Sports Card Show. Charlie Rose Expo Ctr., just off Hwy. 301 S. SH: 9am-4pm, T: 225. Carolina Hobby Expo, PH: 704-786-8373 after 4pm.

Mar 1-2 NC, Charlotte. 5th Annual Toy Collectors Show. Hornets Training Ctr., I-77 "South Carolina" Exit 88. SH: Sat. 9am-4pm, Sun. 10am-4pm, T: 320. Carolina Hobby Expo, PH: 704-786-8373 after 4pm.

Mar 22 NC, Clemmons. 3rd Annual Doll Show. Ramada Inn, I-40, Exit 184. SH: 10am-4pm, T: 120. Carolina Hobby Expo, PH: 704-786-8373 after 4pm.

Apr 12 NC, Asheville. 11th Annual Western Carolina Toy Collectors Show. Nat'l. Guard Armory, I-40, Exit 47. SH: 9am-3pm, T: 125. Carolina Hobby Expo, PH: 704-786-8373 after 4pm.

Apr 19-20 NC, Concord. 8th Annual Toy & Hobby Show. Cabarrus Cty. Fairgrounds, Hwy. 29 & Old Charlotte Rd. SH: Sat. 9am-4pm, Sun. 10am-4pm, T: 252. Carolina Hobby Expo, PH: 704-786-8373 after 4pm.

May 3-4 NC, Raleigh. 3rd Annual Doll & Figure Show. State Fairgrounds, just off I-40. SH: Sat. 10am-5pm, Sun. 10am-4pm, T: 200. Carolina Hobby Expo, PH: 704-786-8373 after 4pm.

May 8-11 NC, Charlotte. Dolls, Toys & Trains Toy Show. New Charlotte Conv. Ctr., 501 S. College St. ABLE Promos., 40 N. Moeller St., Binghamton, NY 13901. PH: 607-722-5041 or FAX: 607-797-0220.

May 31 NC, Fayetteville. 20th Annual Sandhills Toy, Hobby & Sports Card Show. Charlie Rose Expo Ctr., just off Hwy. 301 S. SH: 9am-4pm, T: 225. Carolina Hobby Expo, PH: 704-786-8373 after 4pm.

Jul 12-13 NC, Raleigh. 25th Annual North State Toy Collectors Show. State Fairgrounds, just off I-40. SH: Sat. 9am-5pm, Sun. 10am-4:30pm, T: 285. Carolina Hobby Expo, PH: 704-786-8373 after 4pm.

Jul 26-27 NC, Charlotte. 6th Annual Toy Collectors Show. Hornets Training Ctr., I-77 "South Carolina" Exit 88. SH: Sat. 9am-4pm, Sun. 10am-4pm, T: 320. Carolina Hobby Expo, PH: 704-786-8373 after 4pm.

Aug 9 NC, Fayetteville. 21st Annual Sandhills Toy, Hobby & Sports Card Show. Charlie Rose Expo Ctr., just off Hwy. 301 S. SH: 9am-4pm, T: 225. Carolina Hobby Expo, PH: 704-786-8373 after 4pm.

Sep 6-7 NC, Raleigh. 26th Annual North State Toy & Hobby Show. State Fairgrounds, just off I-40. SH: Sat. 9am-5pm, Sun. 10am-4:30pm, T: 285. Carolina Hobby Expo, PH: 704-786-8373 after 4pm.

Sep 14 NC, Raleigh-Durham. "Barbie Madness Mega Show". Sheraton Imperial Hotel, 4700 Emperor Blvd. SH: 10am-4pm. PH: 919-941-5050.

NORTH DAKOTA

May 10 ND, Minot. Toy Show. All Seasons Arena on State Fairgrounds. SH: 10am-5pm. Marla Hagen, 2005 Westfield Ave., Minot, ND 58701. PH: 701-838-6855.

OHIO

Jan 4-5 OH, St. Clairsville. Winter Train & Toy Show. Ohio Valley Mall, Exit 218 off I-70, Mall Rd. SH: Sat. 10am-9pm, Sun. 11am-6pm. Classic Train & Exhibition, Glenn Groff, 3385 Brigadoon Circle SW, Dalton, OH 44618. PH: 800-833-0612 after 6pm.

Jan 4-5 OH, Canton. CJ Trains-Centre Train & Toy Show. Canton Centre Mall, 4000 W. Tuscarawas. SH: Sat. 10am-9pm, Sun. 12noon-5pm. Jon Ulbright, 941 Buchholz Dr., Wooster, OH 44691. PH: 330-262-7488 after 6pm.

Jan 5 OH, Columbus. Paper Fair. Veterans Memorial Hall, 300 W. Broad St. SH: 10am-5pm. Columbus Prods., Inc., PO Box 261016, Columbus, OH 43226. PH: 614-781-0070.

Jan 12 OH, Sheffield Lake. Cards, Comics & Collectibles Show. Community Ctr., Lake Rd., 2 mi. W. of Rt. 307. SH: 10am-4pm, T: 32-6'. Janet Pleska, 68 Maple Ave., Wakeman, OH 44889. PH: 216-839-2737 or FAX: 216-839-5138.

Jan 19 OH, Dayton-Englewood. Collectible Racing & Hot Wheels Expo. Holiday Inn-Englewood, I-70 at Exit 19. SH: 9am-3pm, T: 40. Redline Racing, 10250 Cedarwood Dr., Union, KY 41091. PH: 606-384-3854.

Jan 19 OH, Akron. Starting Lineup Only Show. Holiday Inn, I-77, Exit 120, Arlington Rd. SH: 10am-4pm, T: 25. Chris Sulyma, 1200 East Ave., Elyria, OH 44035. PH: 216-323-6301.

Jan 26 OH, Mansfield. Toy & Collectible Show. Richland Cty. Fairgrounds, Trimble Rd. Exit off US 30. SH: 10am-4pm, T: 175. Kevin Spore, PO Box 9014, Lexington, OH 44904. PH: 419-756-3904.

Feb 1 OH, Zanesville. Toys, CD & Record Show. Secrest Auditorium, 334 Shinnick St., I-70, Exit 154. SH: 10am-5pm, T: 8'. Ross Noggle, 332 W. Main St., Greenville, OH 45331. PH: 513-548-8521.

Feb 8-9 OH, Columbus. Greenberg's Great Train, Dollhouse & Toy Show. Franklin Cty. Veterans Memorial, 300 W. Broad St. SH: Sat. 11am-5pm, Sun. 11am-4pm. Greenberg Shows, 7566 Main St., Sykesville, MD 21784. PH: 410-795-7447.

Feb 9 OH, Cincinnati-Ft. Mitchell, KY. Collectible Miniatures Show. Drawbridge Inn Estates. SH: 10am-4:30pm. Bright Star Promos., Inc., 3428 Hillvale Rd., Louisville, KY 40241. PH: 502-423-STAR.

Feb 9 OH, Sheffield Lake. Cards, Comics & Collectibles Show. Community Ctr., Lake Rd., 2 mi. W. of Rt. 307. SH: 10am-4pm, T: 32-6'. Janet Pleska, 68 Maple Ave., Wakeman, OH 44889. PH: 216-839-2737 or FAX: 216-839-5138.

Feb 9 OH, Elyria. Starting Lineup Only Show. Holiday Inn, OH Tpke. Exit 8 or I-90 Elyria Exit. SH: 10am-4pm, T: 30. Chris Sulyma, 1200 East Ave., Elyria, OH 44035. PH: 216-323-6301.

Feb 23 OH, Tiffin. Toys & Collectibles Show. Mall, 870 W. Market St., SR 18. SH: 12noon-5pm. Karen Kraft, 870 W. Market St., Tiffin, OH 44883. PH: 419-447-1866.

Feb 23 OH, Richfield. "Tickled Pink" (Featuring Barbie Doll & Friends)! Holiday Inn, 4742 Brecksville Rd. SH: 10am-4pm, T: 100. Joanne Edwards, 586 Crestwood, Wadsworth, OH 44281. PH: 330-334-5900.

Mar 2 OH, Cincinnati. "Barbie Madness Mega Show". Marriott, 11320 Chester Rd. SH: 10am-4pm. PH: 513-772-1720.

Mar 9 OH, Wooster. CJ Trains-Spring Train & Toy Show. Agricultural Research & Development Ctr., Fisher Auditorium, just S. of Wooster on SR 302. SH: 10am-4pm, T: 8'. Jon Ulbright, 941 Buchholz Dr., Wooster, OH 44691. PH: 330-262-7488 after 6pm.

Mar 9 OH, Sheffield Lake. Cards, Comics & Collectibles Show. Community Ctr., Lake Rd., 2 mi. W. of Rt. 307. SH: 10am-4pm, T: 32-6'. Janet Pleska, 68 Maple Ave., Wakeman, OH 44889. PH: 216-839-2737 or FAX: 216-839-5138.

Mar 9 OH, Westlake. Starting Lineup Only Show. Holiday Inn, I-90 Crocker Bassett Exit. SH: 10am-4pm, T: 35. Chris Sulyma, 1200 East Ave., Elyria, OH 44035. PH: 216-323-6301.

Mar 23 OH, Columbus. Annual Doll Show. Aladdin Temple, 3850 Stelzer Rd. SH: 10am-4pm. Vivian Ashbaugh, PO Box 579, Pataskala, OH 43062. PH: 614-587-4722.

Mar 23 OH, Eastlake. Starting Lineup Only Show. Clarion Hotel, Rt. 2 Eastlake Exit. SH: 10am-4pm, T: 25. Chris Sulyma, 1200 East Ave., Elyria, OH 44035. PH: 216-323-6301.

Apr 5-6 OH, Niles. Greenberg's Great Train, Dollhouse & Toy Show. Eastwood Expo Ctr., Rts. 422 & 46. SH: Sat. 11am-5pm, Sun. 11am-4pm. Greenberg Shows, 7566 Main St., Sykesville, MD 21784. PH: 410-795-7447.

Apr 6 OH, Cincinnati. "Cincinnati Loves Barbie" Show. Holiday Inn North, 2235 Sharon Rd. (I-75, Exit 15). SH: 10am-4pm, T: 70. Margie Schultz, PO Box 9371, Cincinnati, OH 45209. PH: 513-321-5260.

Apr 6 OH, New Philadelphia. CJ Trains-Spring Train & Toy Show. Holiday Inn Conference Ctr., 131 Bluebell Dr. SW, I-77, Exit 81. SH: 10am-4pm, T: 8'. Jon Ulbright, 941 Buchholz Dr., Wooster, OH 44691. PH: 330-262-7488 after 6pm.

Apr 6 OH, Kenton. Collector Toy Show. Hardin Cty. Fairgrounds, Fairgrounds Rd. SH: 9am-3pm, T: 65. Lowell Morningstar, PO Box 32, Cable, OH 43009. PH: 937-826-4201.

Apr 13 OH, Mansfield. Toy & Collectible Show. Richland Cty. Fairgrounds, Trimble Rd. Exit off US 30. SH: 10am-4pm, T: 175. Kevin Spore, PO Box 9014, Lexington, OH 44904. PH: 419-756-3904.

Apr 13 OH, Cincinnati-Ft. Mitchell, KY. Teddy Bear Show. Drawbridge Inn Estates. SH: 10am-4:30pm. Bright Star Promos., Inc., 3428 Hillvale Rd., Louisville, KY 40241. PH: 502-423-STAR.

Apr 13 OH, Sheffield Lake. Cards, Comics & Collectibles Show. Community Ctr., Lake Rd., 2 mi. W. of Rt. 307. SH: 10am-4pm, T: 32-6'. Janet Pleska, 68 Maple Ave., Wakeman, OH 44889. PH: 216-839-2737 or FAX: 216-839-5138.

Apr 13 OH, Urbana. 11th Annual Collectors Toy & Craft Show. Champaign Cty. Fairgrounds, Park Ave. SH: 9am-3pm, T: 150. Lowell Morningstar, PO Box 32, Cable, OH 43009. PH: 937-826-4201.

Apr 13 OH, Strongville. Starting Lineup Only Show. Holiday Inn, Select. SH: 10am-4pm, T: 30. Chris Sulyma, 1200 East Ave., Elyria, OH 44035. PH: 216-323-6301.

Apr 20 OH, Akron. Starting Lineup Only Show. Holiday Inn, I-77, Exit 120, Arlington Rd. SH: 10am-4pm, T: 25. Chris Sulyma, 1200 East Ave., Elyria, OH 44035. PH: 216-323-6301.

Apr 27 OH, Columbus. Doll Show. Aladdin Temple, 3850 Stelzer Rd. SH: 10am-5pm. Henrietta Pfeifer, 700 Winchester Pike, Canal Winchester, OH 43110. PH: 614-837-5573.

May 3 OH, Delaware. Hot Wheels, Racing & Collector Toy Show. Fairgrounds, Jr. Fair Bldg., Pennsylvania Ave. SH: 9am-3pm, T: 65. Jack Seitter, 85 Westland Way, Delaware, OH 43015. PH: 614-362-8652.

May 4 OH, Toledo. Collectible Miniature Show. Holiday Inn French Quarter. SH: 10am-4:30pm. Bright Star Promos., Inc., 3428 Hillvale Rd., Louisville, KY 40241. PH: 502-423-STAR.

May 4 OH, Delaware. 7th Annual Farm Toy Show. Fairgrounds, Jr. Fair Bldg., Pennsylvania Ave. SH: 9am-3pm, T: 65. Jack Seitter, 85 Westland Way, Delaware, OH 43015. PH: 614-362-8652.

May 4 OH, Elyria. Starting Lineup Only Show. Holiday Inn, OH Tpke. Exit 8 or I-90 Elyria Exit. SH: 10am-4pm, T: 30. Chris Sulyma, 1200 East Ave., Elyria, OH 44035. PH: 216-323-6301.

May 18 OH, Strongsville. "Barbie Goes To Cleveland". Holiday Inn, 171 & St. Rt. 82. SH: 10am-4pm. Marl, PH: 941-751-6275 or Joe, PH: 213-953-6490.

May 18 OH, Sheffield Lake. Cards, Comics & Collectibles Show. Community Ctr., Lake Rd., 2 mi. W. of Rt. 307. SH: 10am-4pm, T: 32-6'. Janet Pleska, 68 Maple Ave., Wakeman, OH 44889. PH: 216-839-2737 or FAX: 216-839-5138.

May 18 OH, Columbus. Toys, Comic Book & All Childhood Collectibles. Aladdin Temple, Shrine Mosque, Exit 32 (Morse Rd.) off I-270, N. of I-70. SH: 10am-4pm, T: 410-8'. K.C.P. Inc., 809 Beulah Church Rd., Apollo, PA 15613. PH: 412-697-5510.

May 25 OH, Cleveland. Starting Lineup Only Show. Holiday Inn, I-71 W. 150th. SH: 10am-4pm, T: 30. Chris Sulyma, 1200 East Ave., Elyria, OH 44035. PH: 216-323-6301.

Jun 1 OH, Columbus. Paper Fair. Veterans Memorial Hall, 300 W. Broad St. SH: 10am-5pm. Columbus Prods., Inc., PO Box 261016, Columbus, OH 43226. PH: 614-781-0070.

Jun 8 OH, Sheffield Lake. Cards, Comics & Collectibles Show. Community Ctr., Lake Rd., 2 mi. W. of Rt. 307. SH: 10am-4pm, T: 32-6'. Janet Pleska, 68 Maple Ave., Wakeman, OH 44889. PH: 216-839-2737 or FAX: 216-839-5138.

Jun 14-15 OH, Bellefontaine. Collector Toy Show. Logan Cty. Fairgrounds, Lake Ave. SH: Sat. 10am-3pm, Sun. 9am-3pm, T: 55. Lowell Morningstar, PO Box 32, Cable, OH 43009. PH: 937-826-4201.

Jul 12-13 OH, Marysville. Collector Toy Show. American Legion Hall, W. 5th St. SH: Sat. 10am-3pm, Sun. 9am-3pm, T: 42. Lowell Morningstar, PO Box 32, Cable, OH 43009. PH: 937-826-4201.

Jul 13 OH, Sheffield Lake. Card, Comic & Collectibles Show. Community Ctr., Lake Rd., 2 mi. W. of Rt. 307. SH: 10am-4pm, T: 32-6'. Janet Pleska, 68 Maple Ave., Wakeman, OH 44889. PH: 216-839-2737 or FAX: 216-839-5138.

Aug 2-3 OH, Mt. Gilead. Farm Toy Show with 8th Annual Farm Days. Morrow County Fairgrounds, State Rt. 42. SH: Sat. 9am-5pm, Sun. 9am-4pm, T: 45. Jack Seitter, 85 Westland Way, Delaware, OH 43015. PH: 614-362-8652.

Aug 10 OH, Sheffield Lake. Cards, Comics & Collectibles Show. Community Ctr., Lake Rd., 2 mi. W. of Rt. 307. SH: 10am-4pm, T: 32-6'. Janet Pleska, 68 Maple Ave., Wakeman, OH 44889. PH: 216-839-2737 or FAX: 216-839-5138.

Aug 10 OH, Springfield. Central OH Collector Toy Show. Clark Cty. Fairgrounds, Rt. 41. SH: 9am-3pm, T: 165. Lowell Morningstar, PO Box 32, Cable, OH 43009. PH: 937-826-4201.

Aug 17 OH, Tiffin. Toys & Collectibles Show. Mall, SR 18, 870 W. Market St. SH: 12noon-5pm. Karen Kraft, 870 W. Market St., Tiffin, OH 44883. PH: 419-447-1866.

Sep 7 OH, Columbus. Toys, Comic Books & All Childhood Collectibles Show. Aladdin Temple Shrine Mosque, Exit 32 (Morse Rd.) off I-270, N. of I-70. SH: 10am-4pm, T: 410-8'. K.C.P. Inc., 809 Beulah Church Rd., Apollo, PA 15613. PH: 412-697-5510.

Sep 14 OH, Sheffield Lake. Cards, Comics & Collectibles Show. Community Ctr., Lake Rd., 2 mi. W. of Rt. 307. SH: 10am-4pm, T: 32-6'. Janet Pleska, 68 Maple Ave., Wakeman, OH 44889. PH: 216-839-2737 or FAX: 216-839-5138.

Oct 5 OH, Van Wert. Farm Toy Show. Cty. Fairgrounds, Rt. 127. SH: 9am-3pm, T: 75. Lowell Morningstar, PO Box 32, Cable, OH 43009. PH: 937-826-4201.

Oct 11-12 OH, London. Mid-Ohio Farm Toy Show. Madison Cty. Fairgrounds, Elm St. SH: Sat. 2pm-6pm, Sun. 8:30am-3pm, T: 65. Lowell Morningstar, PO Box 32, Cable, OH 43009. PH: 937-826-4201.

Oct 12 OH, Sheffield Lake. Cards, Comics & Collectibles Show. Community Ctr., Lake Rd., 2 mi. W. of Rt. 307. SH: 10am-4pm, T: 32-6'. Janet Pleska, 68 Maple Ave., Wakeman, OH 44889. PH: 216-839-2737 or FAX: 216-839-5138.

Nov 9 OH, Sheffield Lake. Cards, Comics & Collectibles Show. Community Ctr., Lake Rd., 2 mi. W. of Rt. 307. SH: 10am-4pm, T: 32-6'. Janet Pleska, 68 Maple Ave., Wakeman, OH 44889. PH: 216-839-2737 or FAX: 216-839 5138.

Dec 7 OH, Sheffield Lake. Cards, Comics & Collectibles Show. Community Ctr., Lake Rd., 2 mi. W. of Rt. 307. SH: 10am-4pm, T: 32-6'. Janet Pleska, 68 Maple Ave., Wakeman, OH 44889. PH: 216-839-2737 or FAX: 216-839-5138.

OKLAHOMA

Jan 18 OK, Tulsa. Toy & Comic Book Show. Berryhill Community Ctr., 2900 S. 65th W. Ave. SH: 9am-6pm. Darrell Herman, 5303 W. 39th St., Tulsa, OK 74107. PH: 918-446-6129.

Mar 29 OK, Edmond. 13th Annual Toy & Farm Toy Show. Methodist Church Christian Activities Center, corner of Hurd & Jackson. SH: 9am-3pm, T: 90. Terry Keil, 1108 Bank Side Circle, Edmond, OK 73003. PH: 405-348-1830.

Nov 8 OK, Jenks. Teddy Bear Convention. Physical Education Gym, Jenks Central Campus, 205 East B St. SH: 10am-4pm, T: 100. Monica Murray, PO Box 728, Jenks, OK 74037. PH: 918-299-5416.

PENNSYLVANIA

Jan 5 PA, Gilbertsville. Train-O-Rama & Toy Show. Fire House, Rt. 73 (E. of Rt. 100). SH: 9am-2pm. Mary Preudhomme, 233 Long Lane Rd., Boyertown, PA 19512. PH: 610-367-7857.

Jan 12 PA, New Hope. Toy & Train Show. Eagle Fire Co., Rt. 202 & Sugan Rd. SH: 8am-1pm, T: 100. Fred Dauncey, PO Box 222, S. Plainfield, NJ 07080. PH: 908-755-0346.

Jan 12 PA, Trevose. Philadelphia Int'l Antique Toy Convention. Raddison Hotel, PA Tpke. Exit 28 & US 1. SH: 10am-4pm, T: 200. Bob Bostoff, 331 Cochran Pl., Valley Stream, NJ 11581. PH: 516-791-4858.

Jan 19 PA, Leesport. Old Toy & Collector Toy Show, Sports Cards & Racing Memorbilia. Farmers Market Social Hall, Rt. 61. SH: 9am-2pm, T: 150. Charles Gallagher, PH: 610-562-5559.

Jan 26 PA, Shrewsbury. Super Sunday Collectors Toy Show. Fire Hall. SH: 9am-1:30pm, T: 6'. Joe Golabiewski, PH: 410-592-5854 or Carl Daehnke, PH: 717-764-5411.

Feb 1-2 PA, Monroeville. Greenberg's Great Train, Dollhouse & Toy Show. Pittsburgh Expo Mart, US Bus. Rt. 22. SH: Sat. 11am-5pm, Sun. 11am-4pm. Greenberg Shows, 7566 Main St., Sykesville, MD 21784. PH: 410-795-7447.

Feb 8-9 PA, Ft. Washington. Greenberg's Great Train, Dollhouse & Toy Show. Expo Ctr., off PA Tpke. Exit 26. SH: Sat. 11am-5pm, Sun. 11am-4pm. Greenberg Shows, 7566 Main St., Sykesville, MD 21784. PH: 410-795-7447.

Feb 9 PA, New Hope. Toy & Train Show. Eagle Fire Co., Rt. 202 & Sugan Rd. SH: 8am-1pm, T: 100. Fred Dauncey, PO Box 222, S. Plainfield, NJ 07080. PH: 908-755-0346.

Feb 16 PA, Gilbertsville. Train-O-Rama & Toy Show. Fire House, Rt. 73 (E. of Rt. 100). SH: 9am-2pm. Mary Preudhomme, 233 Long Lane Rd., Boyertown, PA 19512. PH: 610-367-7857.

Feb 19 PA, Hummelstown. Toy Action. 249 E. Main St. Snow Date: Feb 25. SH: 9am-3pm, T: 120. Stan Wanner, 308 W. 2nd, Hummelstown, PA 17036. PH: 717-566-0056 or 566-8525.

Feb 22-23 PA, Allentown. ATMA Spring Thaw Train Meet. Agricultural Hall, Fairgrounds, 17th & Chew Sts. SH: 9am-3pm, T: 500. Bob House, 1240 Walnut St., Allentown, PA 18102. PH: 610-821-7886.

Feb 28-Mar 1 PA, Merchanicsburg. Annual Cumberland Valley Farm Toy Show & AUCTION. Cumberland Valley Eagle View School, Rt. 11, 4 mi. N. of I-81 & PA Tpke. SH: Fri. 6pm-9pm, Sat. 9am-3pm. PH: 717-776-5762 or 249-8434.

Mar 2 PA, Shrewsbury. 33rd Collectors' Toy Show. Fire Hall. SH: 9am-1:30pm, T: 6'. Joe Golabiewski, PH: 410-592-5854 or Carl Daehnke, PH: 717-764-5411.

Mar 9 PA, New Hope. Toy & Train Show. Eagle Fire Co., Rt. 202 & Sugan Rd. SH: 8am-1pm, T: 100. Fred Dauncey, PO Box 222, S. Plainfield, NJ 07080. PH: 908-755-0346.

Mar 9 PA, Leesport. Old Toy & Collector Toy Show, Sports Cards & Racing Memorbilia. Farmers Market Social Hall, Rt. 61. SH: 9am-2pm, T: 150. Charles Gallagher, PH: 610-562-5559.

Mar 22 PA, Hummelstown. Fire Co. Show. 249 E. Main St. SH: 9am-3pm, T: 120. Stan Wanner, 308 W. 2nd, Hummelstown, PA 17036. PH: 717-566-0056 or 566-8525.

Mar 23 PA, Allentown. Auto & General Toy Show. Merchants Square Mall, 1901 S. 12th & Vultee St., (1 mi. off Exit 18, I-78). SH: 9am-3pm, T: 150-8'. H.N. Sales & Promos., PH: 610-791-3772.

Mar 29 PA, Hatfield. Bucks-Mont Truck & Farm Toy Show. Sanford Alderfer Ctr., 501 Fairground Rd. SH: 8:30am-2pm. Carl Keeler, PO Box 36, Franconia, PA 18924. PH: 215-855-9041 ext. 527.

Apr 6 PA, New Hope. Toy & Train Show. Eagle Fire Co., Rt. 202 & Sugan Rd. SH: 8am-1pm, T: 100. Fred Dauncey, PO Box 222, S. Plainfield, NJ 07080. PH: 908-755-0346.

Apr 6 PA, Gilbertsville. Doll, Toy & Teddy Bear Show. Fire Hall, Rt. 73, 1 mi. E. of Rt. 100, (6 mi. N. of Pottstown). SH: 10am-3:30pm, T: 300. R&S Ent., 34 N. Vintage Rd., Paradise, PA 17562. PH: 717-442-4279 (eves.).

Apr 20 PA, Leesport. Old Toy & Collector Toy Show, Sports Cards & Racing Memorabilia (held in conjunction with outdoor Flea Market). Farmers Market Social Hall, Rt. 61. SH: 8:30am-2pm, T: 150. Charles Gallagher, PH: 610-562-5559.

Apr 20 PA, Pittsburgh. Teddy Bear Show. Marriott Inn Airport. SH: 10am-4:30pm. Bright Star Promos., Inc., 3428 Hillvale Rd., Louisville, KY 40241. PH: 502-423-STAR.

Apr 27 PA, York. Doll, Toy & Teddy Bear Show. Fairgrounds, Old Main Bldg., Rt. 74S of Rt. 30. SH: 10am-3:30pm, T: 300. R&S Ent., 34 N. Vintage Rd., Paradise, PA 17562. PH: 717-442-4279 (eves.).

May 3-4 PA, Pittsburgh. Toys, Comic Books & All Childhood Collectibles. Expo Mart at Monroeville Mall, Bus. Rt. 22, Exit 6, PA Tpke. SH: 10am-4pm, T: 430-6'. K.C.P. Inc., 809 Beulah Church Rd., Apollo, PA 15613. PH: 412-697-5510.

May 4 PA, New Hope. Toy & Train Show. Eagle Fire Co., Rt. 202 & Sugan Rd. SH: 8am-1pm, T: 100. Fred Dauncey, PO Box 222, S. Plainfield, NJ 07080. PH: 908-755-0346.

Jun 1 PA, New Hope. Toy & Train Show. Eagle Fire Co., Rt. 202 & Sugan Rd. SH: 8am-1pm, T: 100. Fred Dauncey, PO Box 222, S. Plainfield, NJ 07080. PH: 908-755-0346.

Jun 8 PA, Philadelphia. "Barbie Goes To Philadelphia". Marriott Airport, Int'l Airport. SH: 10am-4pm. Marl, PH: 941-751-6275 or Joe, PH: 213-953-6490.

Jun 29 PA, Gilbertsville. Doll, Toy & Teddy Bear Show. Fire Hall, Rt. 73, 1 mi. E. of Rt. 100, (6 mi. N. of Pottstown). SH: 10am-3:30pm, T: 300. R&S Ent., 34 N. Vintage Rd., Paradise, PA 17562. PH: 717-442-4279 (eves.).

Jul 6 PA, New Hope. Toy & Train Show. Eagle Fire Co., Rt. 202 & Sugan Rd. SH: 8am-1pm, T: 100. Fred Dauncey, PO Box 222, S. Plainfield, NJ 07080. PH: 908-755-0346.

Jul 6 PA, Leesport. Old Toy & Collector Toy Show, Sports Cards & Racing Memorabilia (held in conjunction with outdoor Flea Market). Farmers Market Social Hall, Rt. 61. SH: 8:30am-2pm, T: 150. Charles Gallagher, PH: 610-562-5559.

Sep 7 PA, New Hope. Toy & Train Show. Eagle Fire Co., Rt. 202 & Sugan Rd. SH: 8am-1pm, T: 100. Fred Dauncey, PO Box 222, S. Plainfield, NJ 07080. PH: 908-755-0346.

Sep 7 PA, Leesport. Old Toy & Collector Toy Show, Sports Cards & Racing Memorabilia (held in conjunction with outdoor Flea Market). Farmers Market Social Hall, Rt. 61. SH: 8:30am-2pm, T: 150. Charles Gallagher, PH: 610-562-5559.

Sep 7 PA, Trevose. Philadelphia Int'l Antique Toy Convention. Raddison Hotel, PA Tpke. Exit 28 & US 1. SH: 10am-4pm, T: 200. Bob Bostoff, 331 Cochran Pl., Valley Stream, NJ 11581. PH: 516-791-4858.

Sep 7 PA, Harrisburg. 7th Annual Central PA Ornament Collectors Club Swap & Sell. Holiday Inn, 4751 Lindle Rd. SH: 9am-3pm, T: 60. Jean Krohn, 3759 Elder Rd., Harrisburg, PA 17111. PH: 717-564-3252.

Oct 5 PA, New Hope. Toy & Train Show. Eagle Fire Co., Rt. 202 & Sugan Rd. SH: 8am-1pm, T: 100. Fred Dauncey, PO Box 222, S. Plainfield, NJ 07080. PH: 908-755-0346.

Oct 5 PA, Gilbertsville. Doll, Toy & Teddy Bear Show. Fire Hall, Rt. 73, 1 mi. E. of Rt. 100, (6 mi. N. of Pottstown). SH: 10am-3:30pm, T: 300. R&S Ent., 34 N. Vintage Rd., Paradise, PA 17562. PH: 717-442-4279 (eves.).

Oct 5 PA, Philadelphia. "Barbie Madness Mega Show". Radisson Hotel in Trevose, 2400 Old Lincoln Rd. SH: 10am-4pm. PH: 215-638-8300.

Nov 2 PA, New Hope. Toy & Train Show. Eagle Fire Co., Rt. 202 & Sugan Rd. SH: 8am-1pm, T: 100. Fred Dauncey, PO Box 222, S. Plainfield, NJ 07080. PH: 908-755-0346.

Nov 2 PA, Leesport. Old Toy & Collector Toy Show, Sports Cards & Racing Memorabilia (held in conjunction with outdoor Flea Market). Farmers Market Social Hall, Rt. 61. SH: 8:30am-2pm, T: 150. Charles Gallagher, PH: 610-562-5559.

Nov 2 PA, York. Doll, Toy & Teddy Bear Show. Fairgrounds, Old Main Bldg., Rt. 74S of Rt. 30. SH: 10am-3:30pm, T: 300. R&S Ent., 34 N. Vintage Rd., Paradise, PA 17562. PH: 717-442-4279 (eves.).

Dec 6-7 PA, Pittsburgh. Toys, Comic Books & All Childhood Collectibles Show. Expo Mart at Monroeville Mall, Business Rt. 22, Exit 6, PA Tpke. SH: 10am-4pm, T: 430-6'. K.C.P. Inc., 809 Beulah Church Rd., Apollo, PA 15613. PH: 412-697-5510.

Dec 14 PA, New Hope. Toy & Train Show. Eagle Fire Co., Rt. 202 & Sugan Rd. SH: 8am-1pm, T: 100. Fred Dauncey, PO Box 222, S. Plainfield, NJ 07080. PH: 908-755-0346.

SOUTH CAROLINA

Mar 1-2 SC, Fort Mill. 5th Annual Toy Collectors Show. Charlotte Hornets Training Ctr., I-77, Exit 88. SH: Sat. 9am-4pm, Sun. 10am-4pm, T: 320. Carolina Hobby Expo, PH: 704-786-8373 after 4pm.

Jul 26-27 SC, Fort Mill. 6th Annual Toy Collectors Show. Charlotte Hornets Training Ctr., I-77, Exit 88. SH: Sat. 9am-4pm, Sun. 10am-4pm, T: 320. Carolina Hobby Expo, PH: 704-786-8373 after 4pm.

SOUTH DAKOTA

Sep 5-6 SD, Rapid City. Black Hills Area Toy Show. Rushmore Plaza Civic Ctr., 444 Mt. Rushmore Rd. N. SH: Fri. 5pm-9pm, Sat. 9am-4pm, T: 135. Ken Charlton, 810 E. Tallent, Rapid City, SD 57701. PH: 605-343-3643.

TENNESSEE

Jan 4 TN, Chattanooga. Antique & Collectible Toy & Doll Show. Conv. Center's South Hall. Southeastern Collectible Expo, 1819 Peeler Rd., Vale, NC 28168. PH: 704-276-1670.

Jan 25-26 TN, Johnson City-Bristol-Kingsport. Toy, Train, Doll & Hobby Show. Appalachian Fairgrounds (indoor show). T: 350. PH/FAX: 423-474-3910.

Apr 26-27 TN, Knoxville. 3rd Annual Smoky Mountain Toy Collectors Show. Chilhowee Park, Jacob Bldg., I-40, Exit 392. SH: Sat. 9am-4pm, Sun. 10am-4pm, T: 320. Carolina Hobby Expo, PH: 704-786-8373 after 4pm.

May 18 TN, Nashville. "Barbie Madness Mega Show". Regal Maxwell House Hotel, 2025 Metrocenter Blvd. SH: 10am-4pm. PH: 615-259-4343.

TEXAS

Feb 1 TX, Dallas-Ft. Worth. Dolls & Collectibles Show. Arlington Community Ctr. at Vandergriff Park on Matlock Rd., 1 mi. No. of I-20. SH: 10am-6pm, T: 130. Robert Maston, 201 Martin, Lancaster, TX 75146. PH: 214-227-0100.

Feb 2 TX, Houston. "Barbie Goes To Houston". Wyndham Greenspoint Hotel, 12400 Greenspoint Dr. SH: 10am-4pm. Marl, PH: 941-751-6275 or Joe, PH: 213-953-6490.

Feb 8 TX, New Braunfels. Hill Country Doll Show. Civic Ctr., 380 S. Sequin St. SH: 9am-4pm. Dorothy Sojourner, 625 Gruene River Dr., New Braunfels, TX 78132. PH: 210-608-0308 office or 625-3245 home.

Feb 9 TX, Houston. Autograph Show. Medallion Hotel, 3000 North Loop West (Hwy. 290 & 610 Loop). SH: 9am-5pm, T: 40. PH: 713-807-9797 after 5:30pm or 723-0296.

Feb 22 TX, Gainesville. 11th Annual North Texas Farm Toy Show. Civic Ctr., 311 S. Weaver. SH: 9am-4pm. Ed Pick, Rt. 2, Box 266, Muenster, TX 76252. PH: 817-759-2876.
Mar 1-2 TX, Austin. Collectors Exposition. Palmer Auditorium, lower level, S. 1st St. & Riverside Dr. SH: Sat. 9am-6pm, Sun. 10am-4pm, T: 200. Sally Wallace, PH: 512-454-9882.
Mar 2 TX, Houston. Teddy Bear Show. Greenspoint Marriott. SH: 10am-4:30pm. Bright Star Promos., Inc., 3428 Hillvale Rd., Louisville, KY 40241. PH: 502-423-STAR.
Mar 15-16 TX, Plano. DFW Sci-Fi & Action Figure Toy Show. Conv. Ctr. (NW corner of Spring Creek & I-75 Central Expwy.). SH: Sat. 10am-5pm, Sun. 10am-4pm, T: 200. Ben Stevens, PH: 214-578-0213.
Mar 15 TX, Grapevine. 12th Annual Toy Show. Grapevine Conv. Ctr., 1209 S. Main St. SH: 9am-4pm. Martha Gragg, 5910 NW 56th, Oklahoma City, OK 73122. PH: 405-789-2934. FAX: 405-787-6873.
Mar 23 TX, Dallas. "Barbie Madness Mega Show". Harvey Hotel Brook Hollow, 7050 Stemmons Fwy. SH: 10am-4pm. PH: 214-630-8500.
Apr 17-20 TX, Houston. National Paper Doll Convention "Kids at Heart". Holiday Inn Select, 2712 SW Freeway at Kirby Dr. SH: 10am-3pm. Melita Howell, 4129 Southerland Rd. Bldg. E, Houston, TX 77092. PH: 713-690-6558 days or 772-8781 eves.
Apr 26 TX, Longview. Diecast & Figure Show. VFW Hall 4002, 401 Ambassador Row. SH: 10am-5pm, T: 50. Bob Thompson, 1122 Brookshire Dr., Tyler, TX 75701. PH: 903-595-6663.
Jun 22 TX, Houston. "Barbie Madness Mega Show". J.W. Marriott, 5150 Westheimer Rd. SH: 10am-4pm. PH: 713-961-1500.
Jun 28-29 TX, Plano. DFW Sci-Fi & Action Figure Toy Show. Conv. Ctr. (NW corner of Spring Creek & I-75 Central Expwy.). SH: Sat. 10am-5pm, Sun. 10am-4pm, T: 250. Ben Stevens, PH: 214-578-0213.
Jul 11-12 TX, Austin. Collectors Exposition. Palmer Auditorium, lower level, S. 1st St. & Riverside Dr. SH: Sat. 9am-6pm, Sun. 10am-4pm, T: 200. Sally Wallace, PH: 512-454-9882.
Jul 27 TX, Houston. Collectible Miniatures Show. Marriott Inn-Greenspoint, I-45 at N. Sam Houston Pky. E. SH: 10am-4:30pm. Bright Star Promos., Inc., 3428 Hillvale Rd., Louisville, KY 40241. PH: 502-423-STAR.
Jul 27 TX, Irving-Dallas. "Barbie Goes Tto Dallas". Holiday Inn DFW Airport North, 4441 Hwy. 114 & Esters Blvd. SH: 10am-4pm. Marl, PH: 941-751-6275 or Joe, PH: 213-953-6490.

VIRGINIA

Jan 4-5 VA, Chantilly. Greenberg's Great Train, Dollhouse & Toy Show. Capital Expo Ctr., Rt. 28 & Willard Rd. SH: Sat. 11am-5pm, Sun. 11am-4pm. Greenberg Shows, 7566 Main St., Sykesville, MD 21784. PH: 410-795-7447.
Jan 11-12 VA, Virginia Beach. Greenberg's Great Train, Dollhouse & Toy Show. Pavilion, I-64 to Rt. 44 East Exit to end. SH: Sat. 11am-5pm, Sun. 11am-4pm. Greenberg Shows, 7566 Main St., Sykesville, MD 21784. PH: 410-795-7447.
Apr 6 VA, Dunn Loring. 54th Capitol Miniature Auto Collectors Club Show. Volunteer Fire House Community Hall, 2148 Gallows Rd. SH: 9:30am-2pm. James William Brostrom, 6632 Cardinal Ln., Annandale, VA 22003. PH: 703-941-0373 or 301-434-6209.
Apr 12-13 VA, Chantilly. Antiques & Collectibles Show. Capitol Expo Ctr. SH: Sat. 10am-7pm, Sun. 10am-4pm, T: 400. Antique Show Prods., 235 S. Main St., Dunfries, VA 22026. PH: 703-221-3743.
Apr 13 VA, Richmond. Old Dominion Dollhouse Miniatures Show. Embassy Suites Ballroom. Molly Cromwell, 4701 Duncan Dr., Annandale, VA 22003. PH: 703-978-5353.
May 18 VA, McLean. Tysons Corner Spring Dollhouse Miniatures Show. Tysons Holiday Inn, Capital Beltway Exit 11B. Molly Cromwell, 4701 Duncan Dr., Annandale, VA 22003. PH: 703-978-5353.
Jun 7 VA, Richmond. Toys, Comic Book & All Childhood Collectibles. The Showplace Expo Center, 3000 Mechanicsville Tpke., Rt. 360 off I-64 E. SH: 10am-4pm, T: 380-8'. K.C.P. Inc., 809 Beulah Church Rd., Apollo, PA 15613. PH: 412-697-5510.
Oct 19 VA, McLean. Tysons Corner Fall Dollhouse Miniatures Show. Tysons Holiday Inn. Molly Cromwell, 4701 Duncan Dr., Annandale, VA 22003. PH: 703-978-5353.

WASHINGTON

Jan 25 WA, Fife. Antique & Collectible Toy Show. Executive Inn. SH: 10am-4pm. Western Washington Harvest of Toys, Dan Marek, 14202 4th Ave. E., Tacoma, WA 98445. PH: 206-537-3172 or Charlie Ostlund, PH: 206-863-6211.
Jun 21 WA, Tacoma. Aviation & Airline Collectibles Expo. Dome. T: 350. Karl Promotions, PO Box 6543, Lynnwood, WA 98036. PH: 206-744-0983 or FAX: 206-745-2032.
Jun 21 WA, Tacoma. Transportation Expo. Dome. T: 350. Karl Promotions, PO Box 6543, Lynnwood, WA 98036. PH: 206-744-0983 or FAX: 206-745-2032.
Aug 9 WA, Tacoma. Doll, Bear & Toy Expo. Dome. T: 350. Karl Promos., PO Box 6543, Lynnwood, WA 98036. PH: 206-744-0983 or FAX: 206-745-2032.

WISCONSIN

Jan 11 WI, Wisconsin Dells. "Child's Play", Doll, Toy & Teddy Bear Miniatures Show. Holiday Inn Convention Center, I-90 Exit 87. SH: 9:30am-4pm. Aloma Dolan, PO Box 31, Sun Prairie, WI 53590. PH: 608-837-4701.
Feb 2 WI, Milwaukee. Orphans In The Attic Doll, Toy & Bear Show. Serb Hall, 5101 W. Oklahoma Ave. SH: 10am-4pm, T: 140. Marge Hansen, N96W20235 County Line Rd., Meno. Falls, WI 53051. PH: 414-255-4465.
Feb 9 WI, Platteville. 6th Annual Model Car & Toy Show & Swap Meet. Golf & Country Club. SH: 9am-3:30pm. Steve Swift, 320 Mound Ave., Belmont, WI 53510. PH: 608-762-5605 or Jeff Richardson, PH: 608-348-6400.
Mar 16 WI, Ripon. 7th Annual Dairyland Farm Toy Show & Country Craft Show. High School Gym & Cafeteria. SH: 9am-3pm. Steve Meyer, PH: 414-748-2223, Ben Downs, PH: 414-685-2705 or Pat Ehrenberg, PH: 414-748-6088.
Apr 6 WI, Milwaukee. "Orphans In The Attic" Doll, Toy & Bear Show. Serb Hall, 5101 W. Oklahoma Ave. SH: 10am-4pm, T: 140. Marge Hansen, N96W20235 County Line Rd., Meno. Falls, WI 53051. PH: 414-255-4465.
Apr 20 WI, Green Bay. "Orphans In The Attic" Doll, Toy & Bear Show. Best Western Midway Hotel, 780 Packer Dr. SH: 10am-4pm, T: 120. Marge Hansen, N96W20235 County Line Rd., Meno. Falls, WI 53051. PH: 414-255-4465.
Apr 27 WI, Appleton. 9th Annual Fox Valley Area Toy Show. Darboy Club, La Cove Room, Jct. of Hwys. N & KK. SH: 9am-3pm. Robert & Kristine Loderbauer, N121 Cty. Hwy. N, Appleton, WI 54915. PH: 414-730-0292.
May 4 WI, Madison. Orphans In The Attic Doll, Toy & Bear Show. Ramada Inn, 3521 Evan Acres Rd. (I-90, Exit 142B). SH: 10am-4pm, T: 150. Marge Hansen, N96W20235 County Line Rd., Meno. Falls, WI 53051. PH: 414-255-4465.
May 4 WI, Waunakee. 11th Annual South Central WI Farm Toy Show. High School, 100 School Dr. SH: 9am-3pm. Jim Becker, 5729 Cty. Road Q, Waunakee, WI 53597. PH: 608-849-5621.

CANADA

Jan 26 ON, Mississauga. 2nd Annual Toy Show. 6650 Hurontario St. SH: 9:30am-3:30pm, T: 60. Claus Gunzel, 525 Highland Rd. W., Ste. 304, Kitchener, ON N2M 1G5. PH: 519-570-3120.
Feb 9 ON, Simcoe. Toy Show. Royal Canadian Legion, 200 West St. SH: 10:30am-4pm, T: 35. Ian Ward, PH: 519-426-8875.

If You're Searching For A Special Item, Advertise For It!